T0260091

Chemical Engineering for Non-Chemical Engineers

Chemical Engineering for Non-Chemical Engineers

Jack Hipple

AIChE
The Global Home of Chemical Engineers

WILEY

For general information on our other products and services or for technical support, please contact our Customer Care Department within the United States at (800) 762-2974, outside the United States at (317) 572-3993 or fax (317) 572-4002.

Wiley also publishes its books in a variety of electronic formats. Some content that appears in print may not be available in electronic formats. For more information about Wiley products, visit our web site at www.wiley.com.

Library of Congress Cataloging-in-Publication Data

Names: Hipple, Jack, 1946–
Title: Chemical engineering for non-chemical engineers / Jack Hipple.
Description: Hoboken : John Wiley & Sons Inc., 2017. | Includes bibliographical references and index.
Identifiers: LCCN 2016041691| ISBN 9781119169581 (cloth) | ISBN 9781119309659 (epub) | ISBN 9781119309635 (ePDF)
Subjects: LCSH: Chemical engineering–Popular works.
Classification: LCC TP155 .H56 2017 | DDC 660–dc23
LC record available at https://lccn.loc.gov/2016041691

Cover Image: jaminwell/gettyimages; yesfoto/gettyimages
Cover design by Wiley

Set in 10/12pt Warnock by SPi Global, Pondicherry, India

MIX
Paper from
responsible sources
FSC
www.fsc.org FSC® C013604

In a career that has spanned almost 50 years, four organizations, and teaching publicly for the past 15 years, I have to ask myself, "How did all this happen?" An excellent education from Carnegie Mellon University was the start. The opportunities, both technical and managerial, at Dow Chemical, the National Center for Manufacturing Sciences, Ansell Edmont, and Cabot were all important. My accidental introduction to Inventive Problem Solving ("TRIZ") was another. The opportunities to teach "Essentials of Chemical Engineering for Non-Chemical Engineers" for the American Institute of Chemical Engineers and to serve in leadership positions in its Management Division and on its national Board of Directors were the final experiences. But what was the common factor? You cannot move a family multiple times and pursue these opportunities without the support and glue that keeps things together and supports your dreams and ambitions. Sincere thanks to my wife, our four daughters, and our wonderful grandchildren for years of love, dedication, and unyielding love and support.

Contents

Preface

Prior to 1908, individuals in the United States who practiced chemistry on an industrial and commercial scale were members of the American Chemical Society. As you might imagine, and as we will discuss in significant detail, the practice of chemistry on a laboratory scale is quite different and offers significantly different challenges than that same chemistry practiced on an industrial scale. In 1908, in Pittsburgh, Pennsylvania, a group of these individuals held the first meeting of the American Institute of Chemical Engineers, an organization I have been proud to be a member of since 1967, upon my graduation from Carnegie Mellon University.

Today what we know as chemical engineering is practiced in every country around the world. Chemical engineers are employed by nearly every Fortune 500 company in the United States and every large industrial organization in the world. Chemical engineers are employed in the global oil, gas, and petrochemical industries, enabling our entire transportation system. The roads we drive on are manufactured from petroleum products or high-temperature chemical processing of sand and aggregate. In government, chemical engineers serve as employees of the Environmental Protection Agency, the Chemical Safety Board, the Department of Homeland Security, and agencies involved with the oversight of the shipment and processing of chemicals and materials by land, sea, and air. Chemical engineers, in the industrial world, work for energy, food, and consumer companies who produce the energy that heats our homes, powers our cars, and provides the myriad of products we use every day (and take for granted) including simple things as toilet paper and food wrap materials used in packaging and protecting goods; construction materials used in our homes, buildings, and roads; and products that protect our homes and food products from pests and spoilage. They develop products and systems used in photography systems, security systems, and sensor systems. They have developed processes to separate air into its individual components, such as nitrogen and oxygen, allowing the prevention of fire hazards, emergency breathing equipment, and frozen foods. There is virtually nothing that we use or interact with in our daily lives that has not been developed, commercialized, and enhanced by chemical engineers and the knowledge base of chemical engineering.

This book is not intended to be an academic textbook in chemical engineering with detailed equations and complicated mathematical models (there are many excellent ones already available), but to be a layperson's text on what chemical engineering is, its basic principles, and provide simple examples of how these principles can be used to estimate and do rough calculations for industrial equipment and applications. This book is an outgrowth of my teaching a course for the American Institute of Chemical Engineers for the past 15 years entitled "Essentials of Chemical Engineering for Non-Chemical Engineers" (http://www.aiche.org/academy/courses/ch710/essentials-chemical-engineering-non-chemical-engineers). This course has been taken by chemical plant technicians and operators, chemists, biologists, EPA lawyers and interviewing accident psychologists, and Department of Homeland Security inspectors, as well as mechanical and other types of engineers who interact with chemical engineers on a daily basis but may not have a fundamental understanding of the data they are asking for and why, how this information is used, or why they may be asked to perform certain functions in a given way or in a specific order. It is also designed for managers of departments in large organizations, small start-ups, and in organizations who supply equipment and assistance to the chemical industry, but who are not personally chemical engineers. It is also directed at those who are responsible for chemical engineering activities but need a more thorough understanding of what they are managing or directing. A final interested group may be students now entering chemical engineering graduate programs coming in from outside the traditional chemical engineering undergraduate curricula and may find a basic overview helpful in their transition. To all of these potential readers, I hope this book provides some basic understanding and value.

The book is divided into chapters, each focused on a particular aspect or unit operation in chemical engineering and an appendix with the following structure:

1) A discussion and review of the topic basics, using illustrations of equipment where possible
2) A list of general discussion questions for your use within your organization
3) A set of multiple-choice questions on the materials in the chapter, with answers and explanations in the appendix to the book

As we walk through the basics of chemical engineering, the brewing of coffee will be used as an everyday example of the application of many of the principles to an operation that most of us do every day without thinking about the technical principles involved. We will also discuss, at the beginning and end of the book, how chemical engineering principles are used in the manufacture of beer.

Additional materials in the appendix include references and links for MSDS and safety-related material as well as commentary on future challenges for chemical engineering.

At the end of each chapter, a list of additional resources, primarily from AIChE's flagship publication, *Chemical Engineering Progress*, is included. The AIChE website is at www.aiche.org.

Appendix I contains commentary regarding future challenges for chemical engineering. Appendix II contains a list of on line sources of information related to each chapter. Appendix III contains the answers to the multiple-choice questions.

Acknowledgments

For the past 15 years, it has been my personal and professional pleasure to have taught a course for the American Institute of Chemical Engineers entitled "Essentials of Chemical Engineering for Non-Chemical Engineers." AIChE allowed me to teach the basics of a profession that has been the core of my professional life to a variety of people that I could never have imagined. These have included chemists, laboratory and process technicians and operators, other types of engineers, managers of chemical engineers with a different background or training, business managers, safety and health professionals from the government and private sector, patent attorneys, equipment vendors, and psychologists. They have often taught me as much as I knew from their hands on experience in particular areas. I also gratefully acknowledge all my professional colleagues at Dow, the National Center for Manufacturing Sciences, Ansell Edmont, and Cabot—all of whom have advanced my knowledge of chemical engineering and its application in a myriad of applications.

AIChE's flagship publication, *Chemical Engineering Progress*, has been a continuing source of ideas, illustrations, and examples, and this publication will be used and cited throughout this book. Other web-based resources are also used for equipment illustrations and these will be cited at the appropriate time.

1

What Is Chemical Engineering?

There are no doubt numerous dictionary definitions of chemical engineering that exist. Any of these could be unique to the environment being discussed, but all of them will involve the following in some way:

1) Technology and skills needed to produce a material on a commercially useful scale that involves the use of chemistry either directly or indirectly. This implies that chemistry is being used at a scale that produces materials used in commercial quantities. This definition would include not only the traditional oil, petrochemical, and bulk or specialty chemicals but also the manufacture of such things as vaccines and nuclear materials, which in many cases may be produced in large quantities, but by a government entity without a profit motive, but one based on the welfare of the general public.

2) Technology and skills needed to study how chemical systems interact with the environment and ecological systems. Chemical engineers serve key roles in government agencies regulating the environment as well as our energy systems. They may also serve in an advisory capacity to government officials regarding energy, environmental, transportation, materials, and consumer policies.

3) The analysis of natural and biological systems, in part to produce artificial organs. From a chemical engineering standpoint, a heart is a pump, a kidney is a filter, and arteries and veins are pipes. In many schools, the combination of chemical engineering principles with aspects of biology is known as biochemical or biomedical engineering.

The curriculum in all college-level chemical engineering schools is not necessarily the same, but they would all include these topics in varying degrees of depth:

1) Thermodynamics. This topic relates to the energy release or consumption during a chemical reaction as well as the basic laws of thermodynamics that

Chemical Engineering for Non-Chemical Engineers, First Edition. Jack Hipple.
© 2017 American Institute of Chemical Engineers, Inc. Published 2017 by John Wiley & Sons, Inc.

are universally studied across all fields of science and engineering. It also involves the study and analysis of the stability of chemical systems and the amount of energy contained within them and the energy released in the formation or decomposition of materials and the conditions under which these changes may occur.

2) Transport Processes. How fast do fluids flow? Under what conditions? What kind of equipment is required to move gases and liquids? How much energy is used? How fast does heat move from a hot fluid to a cold fluid inside a heat exchanger? What properties of the liquids and gases affect this rate? What affects the rate at which different materials mix, equilibrate, and transfer between phases? What gas, liquid, and solid properties are important? How much energy is required? Materials do not equilibrate by themselves. There is always a driving force such as a pressure difference, a temperature difference, or a concentration difference. Chemical engineers study these processes, their rates, and what affects them.

3) Reaction Engineering and Reactive Chemicals. Chemical reaction rates vary a great deal. Some occur almost instantaneously (acid/base reactions), while others may take hours or days (curing of plastic resin systems or curing of concrete). A chemical reaction run in a laboratory beaker may be where things start, but in order to be commercially useful, materials must be produced on a larger scale, frequently in a continuous manner, using commercially available raw materials. These industrially used materials may have different quality and physical characteristics than their laboratory cousins. Since most chemical reactions either involve the generation of heat or require the input of heat, the practical means to do this must be chosen from many possible options, but for an industrial operation with the potential of release of hazardous materials, the backup utility system must be clearly defined. In addition, chemical reaction rates are typically logarithmic, not linear (e.g., as is the case with heat transfer), providing the possibility for runaway chemical reaction. Chemical engineers must design operations and equipment for such conditions.

4) Safety. There is no basic difference in the hazards or properties of a substance such as chlorine gas on any scale. Its odor, color, boiling point, and toxicity do not change from a small laboratory canister or cylinder to a 10 000 gallon tank car or bulk cylinders used in municipal drinking water disinfection. However, the release of such a material from large-volume processes and tanks can have disastrous consequences to surrounding communities and the people living around them. Any large chemical complex has the same concern about the materials it uses, handles, and produces to ensure that its operations have minimal negative effects on the surrounding community and its customers. The incorporation of formal safety and reactive chemicals education within the college chemical engineering curriculum is a fairly recent and positive development. Chemical engineers are heavily involved

not only in designing and communicating emergency plans for their operations but also in assisting the surrounding communities' emergency response systems and procedures, including ensuring that the hazardous nature of materials used and processes are well understood.

5) Unit Operations. This is a unique chemical engineering term relating to the generic types of equipment and processes used in scaling up laboratory chemistry and the practice of chemical engineering. Heat transfer would be an example of such a unit operation. The need to cool, heat, condense, and vaporize materials is universal in chemical and material processing. The equations used to estimate the rate at which heat transfer occurs can be generalized into a simple equation such that Q (amount of energy transferred) is proportional to the temperature difference (ΔT) as well as the physical characteristics of the system in which the heat transfer is occurring (mixing, physical property differences such as density and viscosity). This would be expressed mathematically as $Q = UA\Delta T$. The amount of energy transferred and the temperature difference may be known, but the "coefficient" (frequently represented by the letter U) relating the two may vary considerably. However, this basic equation can be applied to any heat transfer situation. The same thoughts apply to many separation unit operations such as distillation, membrane transport, reverse osmosis membranes, chromatography, and other "mass transfer" unit operations. The rate of mass transfer is proportional to a concentration difference and an empirical constant, which will be affected by physical properties, diffusion rates, and agitation. In many chemical plant operations, there is an overlap in these areas. For example, a distillation column will involve both heat and mass transfer. The same is true for an industrial cooling tower. The last of these general topics is fluid flow. Though there are many types of pumps and compressors, they all operate on the same basic principle that says that the rate of flow is proportional to the pressure differential, the energy supplied, and the physical properties of the liquid or gas. Again, there is an overlap, as any equipment of this type is also using energy and heating up the liquid or gas it is moving. The heat transfer, as well as the fluid transfer, must be considered.

6) Process Design, Economics, and Optimization. There are numerous ways of scaling up a chemical production system. The choice of particular separation processes, transport systems, storage systems, heat transfer equipment, mixing vessels, and their agitation systems can be done in various combinations, which will impact reliability, cost, the way the process is controlled, and the uniformity of the output of the process. "Design optimization" is a term frequently used. The "optimum" design will not be the same for all companies making the same product as their raw materials base, customer requirements, energy costs, geographic location, cost of labor, and other company unique variables will affect the decision as to what is optimum.

Our ability to computerize chemical engineering design calculations has greatly enabled chemical engineers' capabilities to evaluate a large number of options.

7) Process Control. In a laboratory environment where small quantities of materials are made, the control system may be rather rudimentary (i.e., an agitated flask and on/off heating jacket). However, when this same reaction is "scaled up" orders of magnitude and possibly from batch to continuous, the nature of the process control changes dramatically. The continuous production of specification material around the clock has special challenges in that the raw materials (now coming from an industrial supplier and not a reagent chemical bottle) will not be uniform, the parameters of utilities needed to heat and cool will not be uniform, and the external environment will constantly change. Chemical engineers must design a control system that will not only have to react to such changes but also ensure that there are minimal effects on the product quality, the outside environment, and the safety of its employees.

As the field of chemical engineering has expanded, many curricula will also contain specialty courses in such areas as materials science, environmental chemistry, and biological sciences. However, even when these specialty applications are "scaled" to commercial size, the aforementioned basics will always be needed and considered.

What Do Chemical Engineers Do?

With this type of training, a unique combination of chemistry, mechanical engineering, and physics, chemical engineers find their skills used in a variety of ways. The following is certainly not an all-inclusive list but represents a majority of careers and assignments of most chemical engineers:

1) The scale-up of new and modified chemical processes to make new materials or lower cost/less environmentally impactful routes to existing materials. This is most often described as "pilot plants," which typically is a middle step between laboratory chemistry and full-scale production. In some cases this can involve multiple levels of scale-up (10/1, 100/1, etc.) depending upon the risk factor and the knowledge that exists. A newly proposed process that has operating issues or causes safety releases in a laboratory environment is a serious issue. If that same problem occurs on a much larger scale, the consequences can be far more severe, simply due to the amount and scale of materials being inventoried and processed. These consequences can easily include severe injuries and death, large property damage, and exposure of the surrounding community to toxic materials.

2) Design of Processes and Process Equipment. It is rare that the equipment used in a full-scale plant is identical in type to that used in the laboratory or possibly even in the pilot plant. The piping size; the number and type of trays in a distillation tower; the configuration of coils, tubes, and baffles in a heat exchanger; the shape and size of an agitator system; the shape and geometry of a solids hopper; the shape and configuration of a chemical reactor; and the depth of packing in a tower are all examples of such detailed design calculations. In the commercial world, there may be limitations of certain speeds, voltages, and piping specifications that may not match exactly with what may be desired from smaller-scale work. In these cases, the chemical engineer, in collaboration with other engineers, needs to design a system that will achieve the desired goals, but within practical limitations. In many large chemical and petrochemical companies, chemical engineers will become experts in a certain type of process equipment design and focus most of their career in one particular area.

3) Though certainly not unique to the domain of chemical engineers, the design of utility support systems for chemical plant operations is critical. This includes the supply of water for process and emergency cooling, continuity of electrical supply for powered process equipment such as pumps and agitators, and supply of oil, gas, or coal to generate steam and power. Options chosen will certainly be affected not only by economics but also by limitations of a particular manufacturing site. These may include water availability, water and air permit limitations, and the reliability of local public utility supplies.

4) Sales and marketing positions in the chemical, petroleum, and materials industries are frequently filled by chemical engineers. The ability to understand the customer's process may be critical to the ability to sell a material to a customer, especially if it is a new material or requires substantial change in a customer's operation.

5) Safety and environmental positions, both within industry and government, are frequently filled by chemical engineers. In order to write rules and regulations, it is important to understand the basic limitations of chemical processes, laws of thermodynamics, and the limits of measurement capabilities. Regulations and enforcement actions relating to hazardous material transport also require chemical engineering expertise, especially in bulk pipeline, rail car, and truckload shipping.

6) Cost estimates in the chemical and petrochemical areas are also done by chemical engineers in conjunction with mechanical, civil, and instrumentation engineers. With the availability of today's computer horsepower, it is possible to evaluate and compare many possible process options as a function of raw material pricing, geographic location, energy cost, and cost projections. This allows optimum process design and the ability to predict process costs and economics under changing conditions.

7) The supervision of actual chemical plant operations is most often done by chemical engineers. In this role, the understanding of equipment design and performance is critical, but more importantly the management of plant operations to minimize safety incidents and environmental releases, as well as complying with permits under which the plant is allowed to operate. In this role chemical engineers have additional unique responsibilities including labor relations with operating plant personnel as well as the need, in some cases, to interface with the surrounding community in a public communications role.

8) In universities and in advanced laboratories within many large corporations, basic chemical engineering research is done by advanced degreed chemical engineers, many times in association with other disciplines. Examples of such work would include chemical engineering principles used in the design of artificial organs (remember: the heart is a pump and the kidney is a filter), the study of atmospheric diffusion to study the impact of environmental emissions, the design and optimization of process control algorithms, alternative energy sources and processes, and the recovery of energy from waste products in an economical and environmentally acceptable way.

9) Many business and executive management positions, especially in chemical- and material-based companies (both large and start-up), are filled by chemical engineers. This may come from the advancement over time of newly hired engineers based on demonstrated capabilities (including technical, decision making, and people interaction and motivational skills), as well as the need to transfer chemical engineers with one area of management and technical expertise into another needing, but not having those skills.

10) The study of biological systems from a chemical engineering standpoint. This includes not only the previously mentioned human organs such as the heart (pumps) and kidneys (filters) but also absorption and conversion of food ingredients into the human body.

11) Development of System Models. As our basic understanding of chemical and engineering systems has advanced, it has become easier to mathematically model many process systems. This requires the combination of chemical engineering skills with knowledge of mathematical models and software that, in many cases, minimizes the cost of system scale-up and evaluation.

Topics to Be Covered

The remainder of this book will be divided into chapters represented by the various chemical engineering unit operations, following an overview of safety, reactive chemicals, chemistry scale-up, and economics. The following general introductory topics will be included as necessary as each major chemical engineering unit operation is reviewed:

Chapter 2: Safety and Health: The Role and Responsibilities in Chemical Engineering Practice. There is no perfectly safe chemical (people drown in room temperature water). What particular aspects of safety are important in chemical processing and engineering? What are some examples of materials available and used to evaluate hazards and plan for emergency situations? What kind of protective equipment may be required? What particular aspects of chemical safety require special planning and communication? What are some examples of public- and government-required information? How do we decide on necessary protective equipment? In most cases of commercial processes, there is a wealth of safety and health information available, but discipline is required to review and keep up to date with the latest information. The stability of chemical systems to temperature and heat, oxygen, classes of chemicals, and external contamination must also be understood.

Chapter 3: The Concept of Balances. One of the core principles in chemical engineering is the concept of conservation principles. The amount of mass entering a process or a system, over some period of time, must be equal to what comes out. The same is true for energy with the addition or subtraction of energy change in any chemical reaction and also energy related to equipment operation such as pumps and agitators. Momentum, or fluid energy, is another property conserved in any process.

Chapter 4: Stoichiometry, Thermodynamics, Kinetics, Equilibrium, and Reaction Engineering. How a chemical reaction system, a separation system, a mixing system, a fluid transfer system, or a solids handling system is "scaled up" from their laboratory origins is one of the keys to a commercially successful chemical or materials operation. The methods for doing this are, in most cases, not linear extrapolations. If the scale-up of a chemical process involves many different unit operations, whose scale-up methods are different, this presents a unique challenge in the design of a large-scale chemical process. Chemical reactions either require or generate heat. As will be discussed in more detail later, the rates of reactions and the ability to add and remove heat do not follow the same type of mathematical laws, requiring intelligent engineering design decisions to prevent accidents, injuries, and loss of equipment. Many chemicals will not react with each other under ambient conditions, but their potential products are desirable. Catalysts are materials with special surface properties, which, when activated in a particular way, allow chemicals to react at lower temperature or pressure conditions than otherwise required. They also may allow a higher degree of selectivity of products produced.

Chapter 5: Flow Sheets, Diagrams, and Materials of Construction. As a chemical process idea moves from a laboratory concept to full-scale production, it typically moves through various stages. A mini-plant, a small-scale version of the lab process, might be run to test variables such as catalyst life, conversions and yields being steady over time, reproducibility of the product produced, and similar issues. A pilot plant might then be built, which might be a 100× scale of the lab process, but still 1/10th or 100th the size of the full-scale plant. If it is necessary to provide product samples to a customer during this scale-up process, a semi-plant might be built. The prime function of such a unit is to supply product for customer evaluation, but it will certainly provide additional scale-up and design information. A process being scaled up by a company that is already familiar with the general chemistry may skip one or more of these steps, considering the scale-up risk to be minimal. As this scale-up process moves along, flow sheets that describe how the process will operate and how its various process units will interact with each other become more detailed.

An industrial process rarely uses the same type of equipment used in the laboratory, and one of the key differences is in the materials used in the equipment that handles all the process materials. On a large scale, it is neither safe nor practical to use large-scale glass equipment. Glass-lined equipment is an alternative but can be expensive. Decisions on materials of construction are part of this process of scaling up a process. Decisions on materials to be used must be made and involves corrosion rates and products of corrosion, as well as balancing corrosion rates and product contamination with the possible added cost of corrosion-resistant materials.

Chapter 6: Economics and Chemical Engineering. No chemical reaction or process is commercially implemented unless it provides a profit to someone. Many chemical reactions and formulations are proposed that never go beyond laboratory scale. There must be a demand for the material and the function it provides, and the value (price) of the product or service must be greater than the sum of the cost of its raw materials, the cost of the plant to produce the final product, the cost of any necessary and required environmental controls, the cost of final plant site cleanup and/or disposal, the cost of any borrowed funds invested, the cost of research and development related to the product and process, and the profitability demanded by a company and its shareholders. There may also be unique costs involved in the transportation and storage of any particular chemical.

As previously mentioned, the costs and quality of commercially available raw materials will differ significantly from laboratory reagents. In every case, if the quality will be lower, the raw materials will have impurities that are different, the levels of impurities may change with time, and costs of energy systems may vary. Since the construction of a commercial chemical operation may take many years to complete and the science of forecasting all of these variables is never perfect, estimates are made of changes in these inputs and how they would ultimately affect the cost of manufacture. Economics of making a material is also divided into components that are either fixed or variable, meaning that the costs vary directly with the production volume or they are relatively independent of the volume. The ratio of these two characteristics can have a dramatic impact on chemical or material process profitability as a function of volume and business conditions.

Chapter 7: Fluid Flow, Pumps, and Liquid Handling and Gas Handling. This chapter will review the basics of fluid flow including pumps, gas flow, piping systems, and the impact of changes in process conditions. Fluid transport equipment have limitations that must be understood prior to their choice and use. Fluid mixing can affect chemical reaction rates, uniformity of products produced, and energy costs used by various transport systems. Similar to mass and energy balances, fluid energy and momentum are also conserved in any fluid system, and these potential changes must be accounted for.

Chapter 8: Heat Transfer and Heat Exchangers. Since very few chemical reactions are energy neutral, heat must be either supplied or removed. There are many choices in heat transfer equipment as well as choices in how these various types of equipment are configured. Heat transfer systems are used to heat or cool the reaction systems, insulate piping to maintain a given temperature, maintain temperatures in storage systems, condense gases, boil liquids, and melt or freeze solids. The heating or cooling may also be used to control or change physical properties of a liquid or a gas. It may also be possible to use heat generated in one part of a process to utilize in another part of a process.

Chapter 9: Reactive Chemicals Concepts. This chapter, though separate due to its importance, combines aspects of kinetics, reaction engineering, and heat transfer in the analysis of what is commonly known as reactive chemicals. These aspects of engineering scale in the same way and, if not done correctly, can result in serious loss of life and equipment.

Chapter 10: Distillation. This is the most unique unit operation to chemical engineering. Many liquid mixtures, frequently produced from a chemical reaction, must be separated to recover and possibly purify one or more of the components. If there is vapor pressure or volatility difference between the components, the vaporization and condensation of this mixture done multiple times can produce pure products, both of the more volatile and less volatile components. This unit operation is at the heart of the oil and petrochemical industry that produces gasoline, jet fuel, heating oil, and feedstocks for polymer processes. Low temperature (cryogenic) distillation is also the basis for separating ambient air into its individual components of nitrogen, oxygen, and argon—all used in industrial and medical applications.

Chapter 11: Other Separation Processes: Absorption, Stripping, Adsorption, Chromatography, Membranes. Absorption is the unit operation that describes the removal or recovery of a component from a gas stream into a liquid stream. Stripping is the opposite, or the removal or recovery of a component from a liquid into a gas. Both of these unit operations have become more important over time as environmental regulations have decreased the amount of trace materials that can be discharged directly into the air or water. Adsorption is the use of gas/solid interaction to recover a component from a gas or liquid on to the surface of a solid, the fluid discharged, and the material on the solid surface later recovered via a change in pressure or temperature. The principles of adsorption can also be used to optimize the design of catalyst systems mentioned previously. Charcoal "filters" used to purify home drinking water are an example of this unit operation. Ion-exchange resins are often used to "soften" water for home and industrial use.

Some mixtures require more advanced separation techniques. Water desalination is such an example. Due to basic thermodynamic properties, water would prefer to contain salt, if it is present, rather than to be in its pure state. It is necessary to overcome this "natural" state through the use of permeable selective membranes utilizing a pressure differential. This can be a less costly way of producing drinking water from salt water compared to evaporation. Separation of gases (i.e., air into nitrogen and oxygen) into their components can also be done via membrane-based technologies versus cryogenic (below room temperature) distillation.

Chapter 12: Evaporation and Crystallization. Many chemical reactions result in a product dissolved in a process solvent. This can include salts dissolved in water systems. These types of solutions frequently require concentration to deliver a desired product specification or may require removal of a component whose solubility is lesser than the desire product. Heating or cooling such a solution can be used to evaporate or crystallize the solution and change its concentration of the dissolved solid. This unit operation and its principles overlap with heat transfer topic in Chapter 8.

Chapter 13: Liquid–Solids Separation. Filtration is basically the removal of solids from slurry for the purpose of recovering a solid (possibly produced via evaporation or crystallization). The purpose here could be either recovery of a valuable product, now precipitated, or further processing of a more pure liquid. A drip coffee maker is an example of filtration. This unit operation can be enhanced by the use of gravitational forces such as used in a centrifuge. A home washing machine in its spin cycle is an example of this unit operation.

Chapter 14: Drying. Many chemical products, in their final form, are solids as opposed to liquids or gases. The drying of solids (removal of water or a solvent from a filtration process) involves the contacting of the wet solid with heat in some form (direct contact, indirect contact) to remove the residual water or solvent. The degree of dryness needed is a critical factor in engineering design. The setting used in home clothes dryer is an everyday example.

Chapter 15: Solids Handling. The fundamentals of solids handling and storage are seldom included in chemical engineering curricula at the present time. However, the variables that determine how solids transport equipment (screw conveyors, pneumatic conveyors) operate are extremely important from a practical and industrial standpoint. The characteristics of solids and their ability to be transported and stored are far more complicated than liquids and gases and require the determination of additional physical properties to properly design such process units as bins and hoppers, screw conveyors, pneumatic conveyors, and cyclones. There are also some very unique safety concerns in solids handling, often ignored, that result in dust explosions. The caking of solids in a home kitchen storage unit is an everyday example of what can also happen in industrial processes and packaging.

Chapter 16: Tanks, Vessels, and Special Reaction Systems. Though the actual detailed design of structural supports, pressure vessels, and tanks is normally done by mechanical and civil engineers, the design requirements are often set by chemical engineers. Though tanks and vessels can be used to simply store materials for inventory or batch quality control reasons, they are also used as reactors. This can frequently involve mixing of liquids, gases, and solids; heat transfer; as well as pressure, phase, and volume changes.

Chapter 17: Chemical Engineering in Polymer Manufacture and Processing. These are materials produced from the reaction of monomers such as ethylene, styrene, propylene, and butadiene, which have reactive double bonds. When activated by thermal, chemical, or electromagnetic fields, these monomers can react among themselves to produce long chains of very high molecular weight polymers. Different monomers can be reacted together, producing co- and tri-polymers with varying geometrical configurations. This class of materials has both unique processing and handling challenges due to unusual physical properties and the nonuniform distribution of chemical characteristics. They also have unique challenges in blending and compounding to produce final desired product properties such as color and melting characteristics.

Chapter 18: Process Control. All of the unit operations and their integration into a chemical process require the design of a control system that will produce the product desired by the customer. This chapter also covers the aspects of a control system necessary to deal with the safety and reactive chemical issues mentioned previously.

Chapter 19: Beer Brewing Revisited. In follow-up to the first exercise, we will review the brewing of coffee from the standpoint of chemical engineering principles.

There are also appendices to provide additional discussion and reference materials.

Before we start our journey into the various aspects of chemical engineering, let us take a look at the flow sheet showing how beer is manufactured:

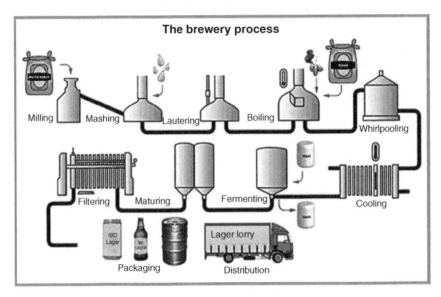

The brewery process

Milling — Mashing — Lautering — Boiling — Whirlpooling

Filtering — Maturing — Fermenting — Cooling

IBD Lager — Packaging — Lager lorry — Distribution

Figure 1.1 Beer manufacturing flow sheet. Source: https://chem409.wikispaces.com/brewing+process. © Wikipedia.

Prior to reading the rest of this book, make a list of some of the chemical engineering issues that you see in designing, running, controlling, and optimizing the brewery process.

1) _____

2) _____

3) _____

4) _____

5) _____

6) _____

7) _____

8) _____

We will revisit this process near the end of this book.

In addition we will use the brewing of coffee (starting at the very beginning) as an illustration of the principles we will present throughout the book.

Discussion Questions

1   What roles do chemical engineers fill in your operations and organization?

2   What unit operations are practiced in your process and facility? Which ones are well understood? Not well understood?

3   How are nonchemical engineers educated prior to their involvement in chemical process operations? Have there been any consequences due to lack of understanding of chemical engineering principles?

4   What chemical process operations are used in your process? How is the knowledge about these unit operations kept up to date? Who is responsible?

5   What areas in your organization's future plans may involve chemical engineering?

Review Questions (Answers in Appendix with Explanations)

1   Chemical engineering is a blend of:
    A __Lab work and textbook study of chemicals
    B __Chemistry, math, and mechanical engineering
    C __Chemical reaction mechanisms and equipment reliability
    D __Computers and equipment to make industrial chemicals

2   Major differences between chemistry and chemical engineering include:
    A __Consequences of safety and quality mistakes
    B __Sophistication of process control
    C __Environmental control and documentation
    D __Dealing with impact of external variables
    E __All of the above

3   A practical issue in large-scale chemical operations not normally seen in shorter-term lab operations is:
    A __Personnel turnover
    B __Personnel protective equipment requirement
    C __Corrosion
    D __Size of offices for engineers versus chemists

4 Issues that complicate large-scale daily chemical plant operations to a much greater degree than laboratory operations include all but which of the following:

A __Weather conditions

B __Emergency shutdown and loss of utility consequences

C __Upstream and/or downstream process interactions

D __Price of company, suppliers, and customer stocks that change minute by minute

5 A chemical engineering unit operation is one *primarily* concerned with:

A __A chemical operation using single-unit binary instructions

B __Physical changes within a chemical process system

C __Operations that perform at the same pace

D __An operation that does one thing at a time

Additional Resources

Felder, R. M. and Rousseau, R. W. *Elementary Principles of Chemical Processes*, 3rd edition, John Wiley & Sons, Inc., 1, 2005.

Himmelblau, D. M. *Basic Principles and Calculations in Chemical Engineering*, Prentice Hall, 1967.

Peters, M. *Elementary Chemical Engineering*, McGraw Hill, 1984.

Solen, J. and Harb, J. *Introduction to Chemical Engineering: Tools for Today and Tomorrow*, 5th Edition, John Wiley & Sons, Inc., 2010.

http://www.aiche.org/academy (accessed August 29, 2016).

2

Safety and Health

The Role and Responsibilities in Chemical Engineering Practice

Basic Health and Safety Information: The Material Safety Data Sheet (MSDS)

We often hear the term "hazardous chemicals" as if this were different from other normal materials. We sometimes describe water as if it is not a chemical, but it is! It has a chemical formula (H_2O), and thousands of people die in it every year (by drowning or being carried away in floods) at room temperature, yet we cannot live without it for more than a few days. Every material has a chemical formula and every material, under some conditions, can cause harm. People are also thermally burned by steam. Gasoline, in conjunction with an internal combustion engine, is a necessary material to move a car, but this same material is highly flammable (but only under certain conditions) and can burn cars to the ground and cause serious burns to a car's passengers, or can be used in arson. It is a chemical engineer's (along with chemists, toxicologists, and biologists) job to clearly define what these hazardous conditions might be and how they might be created and then not only to prevent these situations from occurring but also to communicate clearly to those around us and who work with these materials this same information and the best known ways of dealing with an unsafe situation. For example, what is the best way to put out a gasoline fire? How can it be prevented?

The basic set of information that should be available for any chemical or material includes this information. This listing follows the outline of a particular compound's Material Safety Data Sheet. This is a summary document required to be supplied to a customer of any supplier. A web reference to view an MSDS sheets is included in Appendix II. Very recently, the US Occupational Health and Administration (OSHA) has changed the description of these sheets to simple "SDS," meaning Safety Data Sheets, standardizing on a 15 subtopic format. The link to OSHA's new SDS outline is included in the references at the end of this chapter.

Chemical Engineering for Non-Chemical Engineers, First Edition. Jack Hipple.
© 2017 American Institute of Chemical Engineers, Inc. Published 2017 by John Wiley & Sons, Inc.

1) The Name, Chemical Formula, and Supplier of the Compound. Since some chemicals have "nicknames," which may not give a clear indication of what the material is, this is absolutely essential. The word "water" tells us nothing about what its chemical formula is (though most of us know it from strictly convention and habit), and we could legally describe it as dihydrogen oxide instead of the familiar chemical term "H-2-O." This may seem like a trivial example, but the word "octane" when we hear in conjunction with gasoline can imply several different chemical compounds, all of which contain 8 carbon atoms and 18 hydrogen atoms. There are several different ways of arranging these molecules, and without being specific about the physical arrangement of them (linear? branched? in what way?), we may design an improper fire protection system, supply the wrong octane gasoline, or miscalculate its boiling or freeing point. It might be assumed that any chemical would have the same information regardless of who supplied it, but that is a dangerous assumption. Different suppliers may have different levels of impurities, especially if the same material is produced via a different process, a fairly common situation in the chemical industry. Different suppliers may also have different levels of knowledge about a material they supply.

2) The General (Not Specific) Nature of the Hazards of the Material. Is it particularly toxic? In what form? Is it flammable? Over what range of concentration in air? How easy is it to ignite the material? Are there special hazards to be aware of? Is it water reactive? Is it a strong oxidizer?

3) The Compound's Chemical Composition. As mentioned previously, this needs not only the chemical constituents but also an accurate description of the chemical configuration (more on this later when we discuss geometrical and optical isomers), as well as impurities typically present, to what degree and over what ranges. There is no such thing as a pure material. The impurity level could be at the microscopic level (e.g., PPM) or in the several percent level. As mentioned in #1 earlier, the process to manufacture a chemical could affect the general nature of the impurities.

4) First Aid Measures. If individuals who are involved in the manufacture, distribution, or use of the material were to be exposed to it, what needs to be done? For how long? What do emergency responders need to know? What do emergency room physicians need to know? What special counteractive measures need to be immediately available? To whom? What are the different measures required to deal with inhalation, skin exposure, and oral ingestion? Should vomiting be induced? (You might ask yourself why this is important.) Why or why not? What should be done in the case of accidental spills or releases? What kinds of neutralization or countermeasures need to be immediately available? What kind of protective equipment is needed to handle the material? Gloves, goggles, rubber suits, respiratory masks? Special eye, skin, and respiratory protection? Any special areas of concerns, such as rapid skin absorption or desensitizing olfactory nerves?

5) Fire Hazard and Firefighting Concerns. If the material is totally nonflammable, there would be nothing listed here (in fact "nonflammable" should be clearly stated), but if a material's decomposition under heat or fire could produce flammable or toxic by-products, this would be listed. If a material is flammable, what is its flammability range (the same information reported previously)? What energy levels are required to ignite a material within its flammable range? Does the flammability range change with pressure? There are some materials where water is not the preferred means of extinguishing. For example, a flammable, water-insoluble material may be spread out and make the situation worse. A material may be water reactive. What is the alternative, preferred measure? Is the local fire department aware of this? Do they have the alternative materials (e.g., carbon dioxide) on hand at all times? Are drills held that collaborate with the local emergency responders?

6) Accidental Releases and Spills. What should be done? What kind of emergency prepared response is required? By whom? If a spill could result in a discharge to a water source that is also a municipal drinking water supply, what emergency procedures are in place? Is there frequent and up-to-date communication with municipal authorities and emergency responders? Are there conditions under which corrosion of piping or storage systems could be accelerated? If so, how are they monitored?

7) Handling and Storage. What kind of material should be used to store or transfer the material? Are there special corrosion issues? For example, compounds containing chloride or which may produce chloride ions in a reaction should not be handled in stainless steel. Normal carbon steel may be appropriate. Chloride ions can interact with grain boundaries in stainless steel to produce catastrophic failure, as opposed to just accelerated corrosion. A more expensive material cannot be assumed to be more corrosion resistant. A classic example of this is chlorine's interaction with carbon steel as opposed to the more exotic metal titanium. If chorine is very dry, it can be normally handled in carbon steel; if wet, however, it will aggressively attack steel and rapidly corrode it. Wet chlorine will not attack titanium piping, whereas dry chlorine will react with titanium immediately to form titanium tetrachloride, a reaction that resembles a fire. It is *never* safe to rely on gut feelings to choose materials for storage and piping.

8) Exposure and Personal Protection. We need to know what kind of personal protective equipment (frequently referred to as PPE) is required to be worn by those who handle a given material. If a skin burn or irritation is possible, what kind of PPE is mandated? Gloves? Safety suits? What kind? Made of what material? How often should it be inspected or replaced? Are the potential hazards of materials and the required protection adequately communicated?

9) Physical Properties. All chemicals have defined melting and boiling points, but these can be affected by the presence of impurities. Some materials,

when they melt, can change from a white solid to a colorless liquid, negating the visual sign of the presence of the chemical. Boiling points will vary with pressure. Many times, the boiling point of a material is assumed (from its ambient pressure information) and then not recalculated for a change in pressure, thus allowing vapor escape under conditions not considered.

10) Stability and Reactivity. Many chemicals and materials are stable (meaning that they do not decompose to any measurable degree when stored or used). Others can decompose, to varying degrees, as a function of temperature or in combination with other materials. For example, acids and bases will react with the release of energy. The chemical system that allows a car air bag to protect us (sodium azide) is a chemical that decomposes under shock to generate nitrogen gas which, when released, causes the inflation of the air bags. In this case, we use the known instability for a useful purpose, but obviously, if this reaction occurred without awareness of its occurrence, a major safety incident could result. The type of materials used to contain and transport chemicals is also under this classification. The types of metals and their corrosion rates, as a function of temperature and impurities, need to be known. We previously discussed the unique nature of chlorine systems. The point is that decisions on materials, corrosion, and storage need to be based on data and not assumptions.

11) Toxicology Effects. Different classes of compounds can concentrate in different body organs. For example, halogenated organic compounds tend to concentrate in the liver. Caustic compounds and bases can seriously impact the eyes and skin. It is important to know the toxicological effects of the materials to individuals who handle or may come in contact with the material of concern. This information also needs to be shared with customers and included in an MSDS or SDS sheet supplied to them. This information needs to cover impact and interaction with the skin, eyes, and internal organs such as the lungs and liver. It also usually includes information known to impact unborn children, reproductive organs, and cancer causation.

12) Ecological Information. If released to the environment, how does the material bio-concentrate? In what species? Does it concentrate in sediment? A compound's solubility in water will play an important role in this. How does the compound biodegrade? By what mechanism? What information is required to be supplied to environmental agencies? If a material is released, what is the effect on drinking water supplies?

13) Disposal Information. In laboratory situations, small quantities of chemical wastes may be simply put into special waste containers, and a commercial disposal company removes it. On a larger scale, these types of materials come under a large number of regulatory requirements, which may also

depend on the state in which the operation is being conducted. These requirements must be rigidly adhered to. In some large chemical complexes, there may be ways of recycling or reusing the material. It is also possible to incinerate the material, via combustion or pyrolysis, to generate reusable heat or starting raw materials.

14) Transport Regulations. Many chemicals are strictly regulated by the Department of Transportation (DOT) that regulates how chemicals must be stored, shipped, and labeled. In addition, individual states may have additional labeling and informational requirements. It is critical that chemical engineers, especially those involved in manufacturing and distribution, understand the most up-to-date versions of these regulations.

15) Other Regulatory Information. Information such as "right to know" laws, SARA 313 classifications, MSDS updates, RECRA, etc. are necessary for chemical engineers handling materials to be aware of and ensure compliance. This can be especially important if someone changes geographic locations where the state requirements may be different. There can also be National Fire Protection Association (NFPA) information that must be updated to ensure that DOT shipping labels used to assist emergency responders are up to date.

It is extremely important to keep up to date on MSDS sheets. Though it can become a habit to throw away the latest one received since there are many files already, it is critical to review the latest MSDS sheet and replace older ones with out-of-date information.

Procedures

In addition to basic knowledge of the chemical's properties, it is critical, when handing or processing chemicals, that recognized standard procedures, processes, and safety requirements specified for their use and handling are followed. These are usually above and beyond of what may be included in an MSDS sheet as the way chemicals and materials are handled and processed will vary greatly from company to company and from site to site.

This aspect of chemical handling includes special circumstances such as start-ups, shutdowns, and emergency shutdowns. These procedures are always different than those used in a steady-state operation, and it is difficult to consider all possibilities, requiring awareness and the ability to react to unusual and unanticipated conditions. These can include leaks from tanks and piping, transportation emergencies, emergency reactions to external situations from weather or emergency situations from upstream or downstream processes, loss of utility supplies, and other unanticipated situations.

Fire and Flammability

A particular concern in the chemical and petrochemical industries is handling flammable materials. We are generally familiar with such materials as natural gas, gasoline, butane, propane, and acetylene, but there are many more materials that have potential flammability and explosivity issues. How do we characterize this type of hazard?

Materials that are flammable are not flammable under all conditions. In addition to the chemical (the fuel) itself, two other ingredients are necessary. The first is oxygen. The air in our atmosphere contains approximately 21% by volume oxygen, so oxygen is all around us. Flammable materials require a certain amount of oxygen to burn, but most of the time this level is below 21%, requiring padding with a gas such as nitrogen or argon. The second requirement is heat or an ignition source. The spark plug in our car provides this ignition source in our automobile, igniting a mixture of gasoline and air in certain proportions. This requirement for these three ingredients—fuel, oxygen, and heat/ignition—completes what we know as the fire triangle:

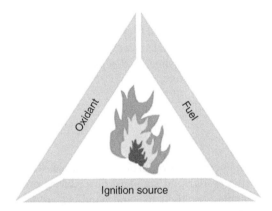

Figure 2.1 The fire triangle. Source: Chemical Engineering Progress, 4/12, pp. 28–33. Reproduced with permission of American Institute of Chemical Engineers.

There are many different color conventions used in this diagram, but the fuel will always be at the bottom of the triangle, the ignition or heat source on the right side of the triangle, and the oxidant on the left side of the triangle. Without all three of these ingredients, it is not possible to sustain a fire. In the context of this diagram, heat is often the by-product of an ignition source. It is also important to remember that an "oxidant" can be something other than oxygen (O_2). Strong oxidizing materials such as peroxides, chlorates, bromates, and iodates can also supply the oxidant function needed. It is easy to assume that the simplest way to prevent a fire is to simply remove heat and ignition sources. While it may be possible to eliminate high temperature sources of heat, it is nearly impossible to eliminate ignition sources. Friction generates heat and is all around us. It is generated by moving machinery parts such as shafts and bearings, friction caused by someone simply walking across a nonconductive surface (as illustrated by the static charge you feel when shaking hands with someone after walking across a carpet in the winter time with low humidity), by something as simple as a fluid flowing through a plastic, non-conductive pipe, or by liquid being sprayed into an open tank. Fires have been caused at traditional gasoline filling stations via the buildup of static filling a plastic versus grounded metal gasoline container. The only proven way of preventing fire is to eliminate the presence of oxygen. This is typically accomplished through the use of inert pad gases such as nitrogen.

We further characterize flammability by what is known as the lower and upper explosive limits (expressed as LEL and UEL, respectively). As flammable as gasoline may be, it is only ignitable and able to sustain a fire within certain limits of oxygen concentrations. At the lower limit, there is too low a fuel/oxygen ratio to sustain a fire. It may be possible to ignite a fire, but the fire cannot be sustained. Blowing out a birthday candle on a cake is an analogous situation as we provide so much air (oxygen) that the fuel/air ratio is too low and the fire goes out. At the other extreme, we can have a situation where the fuel/oxygen ratio is too high and there is not enough oxygen to react with the fuel. The failure of starting a lawn mower engine due to flooding is analogous to this. There is another important property of flammable liquids and that is their autoignition temperature. This is the temperature at which an external ignition source is not necessarily required to initiate combustion. At this elevated temperature, the system has enough thermal energy to act as an ignition source.

LELs and UELs are also affected by pressure and temperature, so measurements of this data must be over the actual planned process conditions. An increase in oxygen concentration, as well as a pressure increase beyond normal atmospheric conditions, can increase the range of flammability.

Minimum ignition energy (MIE) is the amount of energy required to ignite a flammable mixture. This value can vary significantly. For example, the MIE for acetylene is 0.02 mJ, while that for hexane is 0.248 mJ. This makes acetylene far more susceptible to ignition than hexane, but both are extremely flammable. Though ammonia is often considered to be nonflammable, it does have a flammability range between 16 and 25% oxygen. However, ammonia's MIE, is approximately 650 mJ. Though this is several orders of magnitude higher than the more flammable materials and is difficult to ignite, it is NOT nonflammable.

There are two important points about this kind of data. First, it is determined at a given pressure (the information mentioned earlier is at atmospheric pressure). The actual data for the pressure being used in the storage or in the process must be used. Second, the consequences of a fire or explosion are more severe in the middle of the LEL/UEL range. Explosion pressures typically reach their peak in the middle of this range, as shown in Figure 2.2, requiring explosion relief devices to be designed to handle the maximum pressure possible.

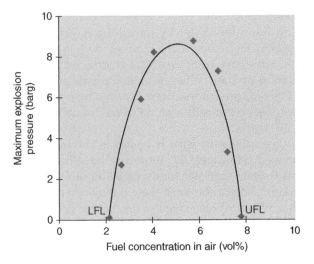

Figure 2.2 Explosion pressure versus fuel concentration. Source: Chemical Engineering Progress, 2/09, pp. 25–29. Reproduced with permission of American Institute of Chemical Engineers.

As with other data of this type, actual laboratory data should be used when available. An additional practical point is that the rate of pressure and energy increase with an explosion. Again, this is a parameter determined experimentally and is important in sizing relief systems. Each flammable material will have its own response curve. It is also important not to mathematically average individual compound data to estimate the explosion pressure for a system but to measure the data for the system as it exists in practice.

We classify explosions into several general categories. Deflagrations are fires and explosions where the propagation of the flame front is less than the velocity of sound (fire or explosion is seen before it is heard). Second, there are detonations, where the propagation of the flame front is less than the velocity of sound (fire or explosion is heard before being seen). "Explosions" describe the rupture of enclosed equipment or piping in an uncontrolled fashion. Fires and explosions can be caused by gases, liquids, or solids and we often ignore the latter. Some of the most deadly fires and explosions have been in the agricultural area with such materials as flour, sugar, and nitrate fertilizers and also in the traditional chemical industry with materials such as plastic dusts that have the capability to build up static charge. The special fire and flammability hazards of solids will be discussed in Chapter 15.

Chemical Reactivity

Classes of reactive chemicals that must be considered in this area for special attention include peroxides, monomers that self-polymerize, oxidizing/reducing combinations, and shock-sensitive materials. In addition, process operations that are heat generating need to be analyzed for their potential to generate sufficient energy to trigger a reactive chemical event. These include mixing and agitation as well as materials of construction that can react with process materials. Another area for review is the handling of inhibited monomers to prevent polymerization during storage and transport.

Special analytical tests, using a technique known as accelerated rate calorimetry (ARC), have been developed for analysis of reactive chemical systems. In these tests a system including the chemicals and materials of concern are slowly heated against a reference standard, and the point at which the reaction or decomposition rate exceeds the system's capability to remove the heat is clearly identified. The video in Appendix II explains the equipment used to determine this information in more detail.

We will discuss the broader subject of reactive chemicals in Chapter 9.

Toxicology

Toxicology is the study of how materials interact with a biological system. Within the scope of this book, we are discussing the effect of chemicals with humans. For many years, the chemical industry has developed and used biological models, primarily based on animal testing, that allow the prediction of toxicological effects of chemicals on the human body as well as specific organs within the body. These tests have been done primarily through the use of mice and rabbits as models for chemical interaction with the human body. Though great progress has been made in identifying methods using cell cultures as an alternative, animal testing still remains the often used method of predicting human organ response to chemical exposure.

There are two general classifications of exposures that we must consider: acute and low level, sustained exposure. For example, the effect of a major spill of an acid, alkali, or toxic gas will be quite different than a slow sustained release of the same material over several years. The first will require a defined emergency response and immediate life-saving procedures, while the second will require constant monitoring and medical testing.

Human exposure to chemicals can occur via several routes, including eye, oral, skin, and lung exposure. A particular chemical may have special greater impact on a certain part of the body or organ, and protection specified against exposure will be affected by this impact. Protective equipment is specified based on our knowledge of these effects. Suppliers of chemicals provide this knowledge to their customers as part of chemical industries' efforts more commonly known as "product stewardship." This concept extends beyond the safe handling of chemicals to include transportation issues and waste disposal and treatment.

Emergency Response

Despite the best plans and intentions, emergency situations can occur due to loss of utilities, human error, and, unfortunately in today's world, terrorism. These types of incidents may occur at the local manufacturing site or thousands of miles away as a result of transportation accidents.

Dealing with emergencies requires preplanning, as well as practiced reaction and drills to unanticipated events. Preparedness for such events includes communication (to employees, emergency responders, and surrounding communities), ensuring that the local community and transportation providers have the most updated safety, toxicological, and medical information. Drills and preplanning are often done in conjunction with local emergency responders and medical facilities that may be called upon to treat victims of unanticipated releases or exposures. In many bulk commodity chemical areas such as chlorine, suppliers have banded together in a pact to allow the nearest manufacturer to respond to an emergency, as opposed to the original supplier having to travel thousands of miles and delay the necessary response.

Transportation Emergencies

When chemicals are transported by land, sea, or air, response to spills or leaks caused by accidents can be a major challenge to emergency response personnel, as they do not have the training or awareness of a local emergency response group close to a chemical facility. For this reason the NFPA, in collaboration with chemical industry, developed a visual diamond symbol for use on transportation vehicles, as seen in Figure 2.3.

Though this diamond typically does not tell the emergency responder the name of chemical formula, it does provide the following important information:

1) What is the nature of the health hazard? Is it relatively harmless or a highly toxic? This section of the diamond is typically shown in a dark gray color (left section of the diamond).
2) What is its flash point? This tells the emergency responders the degree they need to consider serious fire and explosion potential, including keeping spark-producing devices (such as cars) away. This section of the diamond, at the top, is typically a medium gray color.

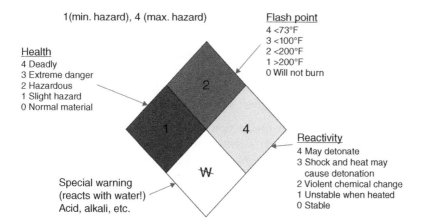

1(min. hazard), 4 (max. hazard)

Flash point
4 <73°F
3 <100°F
2 <200°F
1 >200°F
0 Will not burn

Health
4 Deadly
3 Extreme danger
2 Hazardous
1 Slight hazard
0 Normal material

Reactivity
4 May detonate
3 Shock and heat may
  cause detonation
2 Violent chemical change
1 Unstable when heated
0 Stable

Special warning
(reacts with water!)
Acid, alkali, etc.

Figure 2.3 The NFPA diamond. Source: Reproduced with permission of U.S Department of Transportation.

3) How reactive is the material? Is it shock or heat sensitive? This section of the diamond is typically shown in a light gray color, in the right section of the diamond.
4) The white section of the diamond at the bottom may be blank, but it is the place where any special hazards can be indicated. In this particular case in Figure 2.3, this section shows that the material reacts with water, so the emergency responders might use carbon dioxide (CO_2) as the fire-extinguishing media, as opposed to water. Public emergency responders are generally well educated in this diamond as well as the meaning of the colors, numbers, and symbols within it.

HAZOP

As a result of a major incident at a petroleum depot in Flixborough, England, and others that followed, a formal review process was instituted within the chemical industry that is now known as hazard and operability study (HAZOP). This review process has become one of the bulwark review processes used in the chemical industry and takes a different form than a typical review where the specified flows, temperatures, pressures, and levels are reviewed, and the primary focus is on how to maintain and control the parameters within their specified limits.

In a HAZOP review, we do just the opposite. We ask, in a very systematic way, what happens if the parameters are NOT where they are designed to be.

For example, if a chemical reaction vessel is designed to operate at 100 psig and 200°C, we might ask the following types of questions:

1) What happens if the pressure is lower than 100 psig? Higher than 100 psig? How much higher or lower? For how long?
2) What happens if the reactor temperature is lower than 200°C? Higher than 200°C? How much lower or higher? For how long?

Are the consequences merely production of an off-spec product which may have to be recycled or disposed of? A release of a toxic material? Production of side products contaminating the primary product? Rupture of a vessel? Fire or explosion? Of what magnitude?

Obviously there are many more process variables in addition to temperature and pressure in a process. These would include rates of flow, direction of flow, lack of flow, compositions, sequence and rates of additions in batch processes, potential contamination of desired material flows, direction of flow, substitution of the wrong material, and many others.

We can easily envision a simple household example of this process thinking about our home showers and asking these types of questions, whose answers are straightforward and may depend on a particular situation, just as in a chemical plant:

1) The temperature of the hot water is supposed to be 120°F, controlled by a thermostat in a hot water tank. What if the water comes out of the tank at 140°F? 160°F? Thermal burns are seen typically at 140°F, so this control is critical. How many "backup" controls on the hot water heater are necessary? Desirable? Affordable? Most people using a shower blend the hot and cold water to give a desired temperature. Could 140–160°F water be produced if the supply of cold water is stopped? How could this happen? Does the hot water supply need to be automatically shut off if this happens? During a plumbing repair on a system, could the cold and hot water supplies to the valves be reversed? What happens if the heating unit fails? An uncomfortably cold shower or do the water lines freeze? If they freeze and burst, what are the consequences? To whom? Under what circumstances?
2) The pressure in the water system is controlled to a great degree by the public utility or a privately operated water well system. What if the pressure suddenly drops because of a water main break? Is this a major safety issue or just a major inconvenience? What if it rises unexpectedly? Could a pipe burst? Where would the leaking water go?
3) The plumbing system is normally designed to allow water to drain from the tub or shower. What happens if the drain clogs and the tub overflows? Where does the water go? Is there a difference between the impacts of such an event if the shower is on the 10th floor of a building versus the ground floor? How long does the water need to flow with a plugged drain to cause a problem?

4) Could something other than water get into the supply piping? If so, what could it be and how would it be detected? Can too much water softener cause an extremely slippery surface? Could backup involving waste water happen? How?

The key to these types of reviews is the discipline of asking all of the right questions about all aspects of a process and not to rely on random thought or brainstorming. Table 2.1 shows a list of typical questions used in a HAZOP review and an example of their meaning.

Table 2.1 Typical HAZOP questions.

NO (none of design intent achieved)	Flow is 0 versus 40 GPM design
MORE (more of, higher) quantitative increase in parameter	Flow is 60 versus 40 GPM design
LESS (less of, lower) a decrease in parameter	Flow is 20 versus 40 GPM design
AS WELL AS (an additional activity occurs)	Material is contaminated
PART OF (only part of the design intent is achieved)	40 GPM for 10 min versus design of 40 GPM for 20 min
REVERSE LOGICAL (opposite of design intent occurs)	40 GPM is pumped into a tank versus a design of 40 GPM leaving a tank
COMPLETE SUBSITUTION (vs. design intent)	40 GPM of chemical B versus the intended 40 GPM of chemical A
WHERE ELSE (vs. design intent)	Flow goes to tank A versus tank B
BEFORE/AFTER	In a batch sequence, A is added after B versus the design of A and then B
EARLY/LATE	A is added 15 min into the batch sequence versus 10 or 20 min
FASTER/SLOWER	A is added at 40 GPM 10 min into the batch sequence versus 30 GPM design intent
COUPLING	In addition to a flow error, timing sequence and/or direction of flow is also an error

Many industry accident case studies have taught us valuable lessons that can be applied in this area as well as safety in general:

1) Communication. In large organizations, it is possible for different groups of people to have different levels of knowledge about chemicals, which is not necessarily passed on to all those whose safety may be impacted. Examples include chemical decomposition temperatures, reactivity information, corrosion information, and effects of extreme temperature.
2) Material Stress. Chemical engineers, in general, are not deeply trained in materials science. We often forget that metals change volume with temperature.

This must be considered in the start-up and shutdown of process equipment, especially heat exchangers that may be using high temperature fluids and/or refrigerants. These types of equipment must have defined, orderly, gradual start-up and shutdown procedures to minimize the chances of tubes breaking from tube sheets or high pressure being unexpectedly created by blocking flow of a refrigerant that could evaporate and create pressure.

Layer of Protection Analysis (LOPA)

There is another widely used safety analysis system that can add to conventional checklists, which is the concept of "layers of protection." As a simple illustration, consider leaving your house for a trip. How you approach this analysis may depend on whether you are living alone or with someone you care about. It may also depend on where you live and the attention of your neighbors. Assuming you are worried about a possible intruder, your first "layer" is to lock the doors and set up possibly a motion detector around your house. Depending on your confidence in this first layer and whether your spouse is a sound sleeper, you might jam doors or windows in an attempt to cause a break-in to generate significant noise from glass shattering and arouse someone who is sleeping. If you don't consider this commotion enough to scare away an intruder, you could install a local alarm, triggered by the opening of doors or windows. Whether this alarm is actually hooked up to an emergency call to the local police may be irrelevant if the intruder thinks it is. The sound of the alarm may also alert neighbors who may investigate what is going on. Then of course that alarm can be connected to an actual central alarm station that will call the local police and in addition have a live movie camera attached, which will film the intrusion and provide additional evidence for the police. Now if your house is a mansion full of precious gems, you might add to all of these a hired guard, and just in case you're really worried, you can hire a backup guard.

Each of these layers cost additional money. How do you decide on how much money to spend? That depends. What is the likelihood of the event happening? What is the history of robberies and intrusions in your neighborhood? How soundly does your spouse sleep? How reliable are the various alarm companies and what do they charge? Is it allowable to have a hired gun in your subdivision? There is no one correct answer here, but each of us goes through this type of analysis, even though somewhat superficially, when we leave our house or when we leave our car unattended. Do we leave it open, saving a few minutes when we return? Do we leave the keys in the car? We make a mental calculation of the risk and what we are willing to spend to minimize or eliminate it. We can use this same approach in deciding what level of investment and the nature of that investment that will minimize the impact from a hazardous operation or potential release.

Let's consider the example of a tank containing a hazardous material, either because of its potential fire hazard or the toxicity of the material in the tank. At the start we may have some regulatory requirements to meet such as diking, vent control, and others depending on the nature of the chemical and the state or county in which the tank is located. But after this we have decisions to make. We need to know the level in the tank to ensure that it does not overfill. Is one level indicator enough? What does an additional one cost? What are the financial, environmental, and business implications of the tank overflowing? How reliable is the instrument itself? What is the reliability of the utilities that supply the energy to run it? Do we need a backup air or electricity supply? Is the tank a raw material supply to another part of the same complex? Another customer? Are the consequences the same? There are no right or wrong answers to any of these questions, but they must be consciously answered, and decisions must be made as to the cost of measurement, assessment, and reaction for each system we have concerns about. We also have to think seriously about all of the possible "initiating" events that might trigger a response and how often they might occur. These last two items can change with time and need to be revisited on a regular basis.

An LOPA for a process reactor can be seen in Figure 2.4. An initiating event (i.e., pipe leak, spark, incorrect recipe for a batch reaction, high pressure, or loss of electrical power) may trigger an "event." For example, a pipe leak can happen due to an overpressure or corrosion problem. In our initial design, we may have incorporated a leak detection system for such a possibility tied to a control system to shut off flow through this pipe (levels 2, 3, and 4). The second layer could very well duplicate instrumentation and shut off systems that are activated if the first two do not function. Is the pipe leak in a cooling water supply to an exothermic reactor which may, if cooling is lost, cause a runaway and release materials in an uncontrollable way? This layer may have embedded within it a way for each instrument to communicate with each other as to proper functioning. That may be adequate. Is a third layer necessary? It still depends. On what? How close is the operation to the surrounding residential area? The science of LOPA calculations can become quite complex, and software programs and training is available to assist in these efforts. It is worthwhile noting that the only way to never have a release of any sort is not to build a plant.

Summary

The safe handling of chemical and the safe operation of the plants that make them is a primary responsibility of a chemical engineer. Many informational, evaluation, and analytical tools have been developed over the years to ensure that we have the most appropriate tools and data we need to provide this assurance to the organizations and communities we work with.

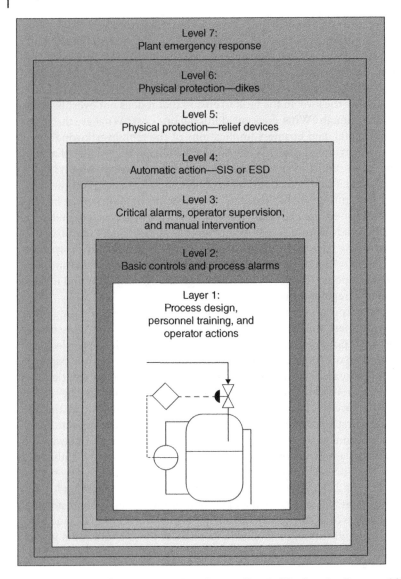

Figure 2.4 A layer of protection analysis. Source: Chemical Engineering Progress, 2/16, pp. 22–25. Reproduced with permission of American Institute of Chemical Engineers.

The American Institute of Chemical Engineers (AIChE) incorporates as part of its professional ethics statement, one which reads as follows:

Engineers shall hold paramount the safety, health and welfare of the public in the performance of their duties

This is what everyone involved in chemical processing needs to remember and *never* lose sight of.

Coffee Brewing: Safety

What safety aspects are involved in this chemical process? Isn't it a chemical process? Does coffee sitting on a hot plate for several hours taste the same as when it first was dripped or perked? There is a chemical degradation reaction occurring. Whether we are brewing instant coffee or producing percolator or drip coffee, hot water is needed. Where does it come from? Depending on your preferences, you might use hot water from a spigot to add to instant coffee in a cup. This is usually not hot enough to produce a desirable beverage, so the water may come from an electrically heated water stream in the sink. Since water can scald (cause first degree burns above 140°F), we should ask how this setting is made. What can cause it to fail? Inject too much electrical heat into the water? Since we are using water around an electrically powered machine (coffee brewer or from a chemical engineering standpoint, a leaching process), can a short develop and possibly electrocute the user? We see ground fault interrupters (GFIs), usually required by building codes (safety rules and procedures), interrupt the supply of electrical current faster than anyone can do physically and, in most cases, prevent a shock or electrocution. We will learn more as we move further along in the course.

Discussion Questions

1   What kinds of fire, health, and environmental impacts are present in your processes? Is there a standard review process? How often is it done? Who attends? What is done with the information? Filed away or brought out for constant review?

2   After any incidents relating to fire, safety, health, or environmental discharges, what kind of reviews are done? Is it established why the normal protective processes did not work? If procedures were in place, why were they not followed? What kind of disciplinary or training changes are needed? Have there been any changes in the surrounding community that would warrant a revisiting of past practices?

3   Are the elements of the fire triangle well understood? Is the assumption of no ignition source being present a method of prevention? If inadequate oxygen is relied upon as the primary prevention mechanism, how is it checked? What could circumvent it? How many layers of protection are

needed? How was the number of layers decided? What factors were considered in the cost/safety decisions?

4 Have all of the issues and variables in scaling up a lab process been considered? What is the next scale level that is appropriate? How much of the decision was economics versus technical risk? How was this balance decided? What factors were involved in the decision making?

5 Is flammability data for the actual chemical mixtures involved being used or mathematically averaging single comment data? Have the effects of pressure variation been considered? Do you have the actual data? If not, how do you plan to obtain it? When?

6 Are sensitive and reactive chemicals being handled? Are there proper storage and separation conditions? If refrigeration is needed, how many layers of protection are appropriate to ensure unsafe temperatures are not reached? Does the local fire department know how to handle such materials? Do they know what to do if asked for their help? Do they even know that such materials are used?

Review Questions (Answers in Appendix with Explanations)

1 Procedures and protective equipment requirements for handling chemicals include all but:
A __Expiration date on the shipping label
B __MSDS sheet information
C __Flammability and explosivity potential
D __Information on chemical interactions

2 Start-ups and shutdowns are the source of many safety and loss incidents due to:
A __Time pressures
B __Unanticipated operational and/or maintenance conditions
C __Lack of standard procedures for unusual situations
D __All of the above

3 The "fire triangle" describes the necessary elements required to have a fire or explosion. In addition to fuel and oxygen, what is the third item that must be present?
A __Ignition source
B __Lightning

C __Loud noise

D __Shock wave

4   The NFPA "diamond," normally attached to shipping containers, indicates all of the following except:

A   __Degree of flammability hazard

B   __Degree of health hazard

C   __Name of chemical in the container

D   __Degree of reactivity

5   The lower explosive limit (LEL) and upper explosive limit (UEL) tell us:

A   __The range of flammability under some conditions

B   __The range of flammability under all conditions

C   __The upper and lower limits of the company's tolerance for losses

D   __The upper and lower limits of the amount of flammable material pumped into a vessel

6   Autoignition temperature is the temperature at which:

A   __The material loses its temper

B   __A material automatically explodes

C   __A material, within its explosive range, can ignite without an external ignition source

D   __Fire and hazard insurance rates automatically increase

7   Toxicology studies tell us all but which of the following:

A   __The difference between acute and long-term exposure effects

B   __Repeated dose toxicity

C   __Areas of most concern for exposure

D   __To what degree they are required and how much they cost

8   An MSDS sheet tells us:

A   __First aid measures

B   __Physical characteristics

C   __Chemical name and manufacturer or distributor

D   __All of the above

9   A HAZOP review asks all of these types of questions except:

A   __Consequences of operating outside design conditions

B   __What happens to the engineer who makes a bad design assumption

C   __Safety impact of operating above design pressure conditions

D   __Environmental impact of discharge of material not intended

Additional Resources

Crowl, D. (2012) "Minimize the Risk of Flammable Materials" *Chemical Engineering Progress*, 4, pp. 28–33.

Fuller, B. (2009) "Managing Transportation Safety and Security Risks" *Chemical Engineering Progress*, 2, pp. 25–29.

Goddard, K. (2007) "Use LOPA to Determine Protective System Requirements" *Chemical Engineering Progress*, 2, pp. 47–51.

Grabinski, C. (2015) "Toxicology 101" *Chemical Engineering Progress*, 11, pp. 31–36.

Karthikeyan, B. (2015) "Moving Process Safety into the Board Room" *Chemical Engineering Progress*, 9, pp. 42–45.

Wahid, A. (2016) "Predicting Incidents with Process Safety Performance Indicators" *Chemical Engineering Progress*, 2, pp. 22–25.

Willey, R. (2012) "Decoding Safety Data Sheets" *Chemical Engineering Progress*, 6, pp. 28–31.

www.nfpa.org (accessed August 30, 2016).

www.nist.gov/fire/fire_behavior.cfm (accessed August 30, 2016).

https://www.osha.gov/Publications/OSHA3514.html (accessed August 30, 2016).

3

The Concept of Balances

Let's explore this general concept in more detail as it underlies so much of fundamental chemical engineering analysis, thinking, and problem solving.

Mass Balance Concepts

Assuming that we have the appropriate understanding of the fundamentals of our reaction and the physical properties of our materials, we can begin to think about the design of a process to manufacture the materials of interest. There are two fundamental concepts that must be understood and taken into account into the design and operation of any process. First, mass must be conserved. Put another way, what comes out of a process must be what we put into it *plus* whatever mass is accumulating within a system.

Chemical Engineering Progress and the Beacon publication from AIChE's Center for Chemical Process Safety have reported many cases of major fire and environmental disasters via the simple mechanism of overflowing tanks. The most significant of these was the Flixborough disaster, summarized in a *Chemical Engineering Progress* "Beacon" article (9/2006, p. 17). Faulty instrumentation, an inadequate number of layers of protection, and insufficient communication have all been causes of major environmental and fire disasters.

All of the process control instrumentation we use in actual product processes are designed to calculate balances but a number of things can happen that can cause surprises:

1) A reaction, with a very slow kinetic rate constant (more later), may be forming a material in the process that was unanticipated. Depending upon the downstream processing and how the anticipated products are handled (distillation, filtration, crystallization, etc.), this material may "bleed" out of the system after building up to a certain level. So eventually the mass balance will close, but in a short or intermediate time frame, it may not.

Chemical Engineering for Non-Chemical Engineers, First Edition. Jack Hipple.

2) An unanticipated liquid reaction product, with a boiling point, which interferes with a distillation separation downstream, can cause a temporary buildup of material.
3) An unanticipated solid reaction product that precipitates out of solution and builds up in process equipment, possibly blocking process piping.

We can consider a very simple example of a material balance for boiling a salt solution to increase its concentration, as shown in Figure 3.1. Let's assume that we have 100# of a 10% salt solution entering an evaporator, and we want to increase its concentration to 50%. How much water has to be evaporated? Let's first start with the feed stream. We have 100 total pounds with 10% salt, so 0.10×100 means there are 10 pounds of salt in the feed stream. That also means there must be $100 - 10$ or 90 pounds of water in the feed. How much salt is in the concentrated product coming out of the evaporator? Since we know that salt does not have a vapor pressure and will not boil, we know that *all* of the salt must be coming out of the evaporator, meaning there is also 10# salt leaving the evaporator. How much water (x) was evaporated? We can calculate this with a simple mass balance calculation. If there is 10# of salt leaving the evaporator (the same as entered) and the solution is 20%, then the amount of water leaving the evaporator can be calculated as follows:

$$10 = 0.20(y),$$

where y is the total solution amount.

$$y = \frac{10}{0.20} = 50 \text{ pounds of total solution}$$

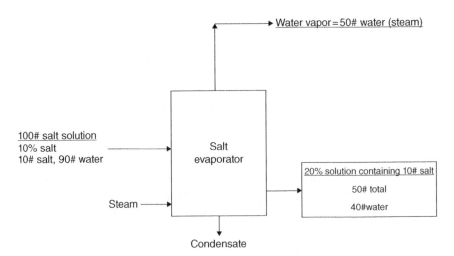

Figure 3.1 Salt evaporator material balance.

Since we know that the solution leaving contains 10# of salt, the remainder must be water, so the water leaving the bottom of the evaporator is 50 – 10 or 40#. How much water is evaporated? The material balance on the water shows us that this must be 90 – 40 or 50# of water. This calculation is shown in Figure 3.1.

If we were measuring one or more of these variables and the balance did not "close," then that indicates a problem that could be any one of these:

1) The steam supply is inadequate to evaporate as much water as desired.
2) The concentration of the incoming salt solution is less than designed.
3) The rate of product leaving the evaporator is not what it is expected to be.
4) The concentration of the salt solution leaving the evaporator is not as desired.

Just the simple concepts of mass balances can generate serious troubleshooting questions to be explored by the people running the evaporator. This may be made easier with any number of online instrumentation systems, but the simple concepts of material balance ("what comes in will come out eventually") can focus the analysis. The accuracy and calibration of instrumentation such as weigh cells, flow meters, and concentration measurements can be critical in determining accurate material balances.

Another important aspect of balances is where we draw the boundaries. Let's take a look at a system where a slurry (a liquid containing suspended solids) is being pumped into a settling tank where solids settle out and the settled solids are drawn off in parallel with the clarified water (see Figure 3.2).

If we draw a box around the entire operation, we would expect that, over time, the amount of slurry entering the tank would equal the amount of water and the amount of enriched or concentrated solids to be equal.

However, we could draw the box at the point shown in Figure 3.3.

In this case, the amount of slurry entering the feed line must equal what actually enters the settling tank. If it did not, there is a leak in the pipe, or some of the solids are accumulating in the pipe, potentially plugging it if left unchecked.

We could also draw the system boundary for a mass balance around the tank itself as shown in Figure 3.4.

In this case, we are measuring the input slurry mass and the mass in the tank. In its initial start-up, we would not expect this mass balance to "close" as the tank will fill up to some predetermined level prior to water overflowing and possibly solids building up to a point where they can be lifted out of the tank.

In Figure 3.5, we see the total system mass balance. If, after some operating time, the sum of the mass of slurry in the tank did not equal the sum of the mass of solids and clarified water leaving the tank, it could indicate:

1) There is a leak in the system in the entering pipe (prior to its entry into the tank), the tank itself, or the clarified water outflow pipe.

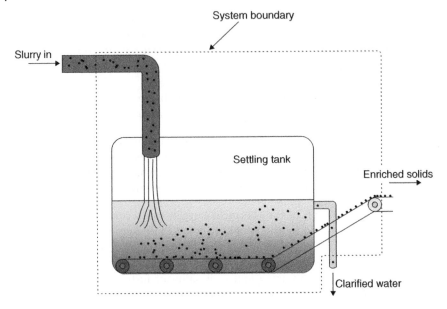

Figure 3.2 Slurry settling and concentration process.

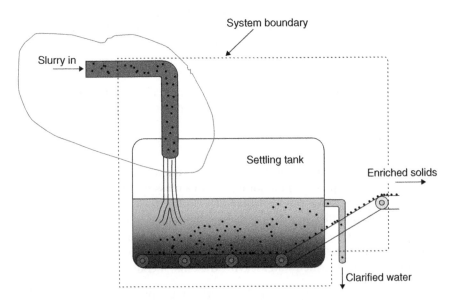

Figure 3.3 Mass balance around feed pipe.

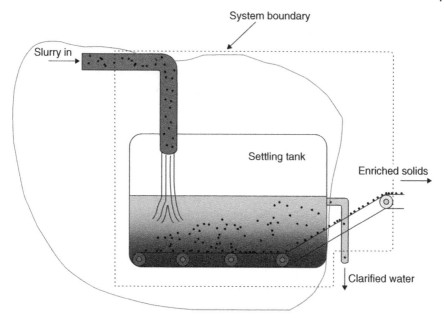

Figure 3.4 Mass balance around the tank alone.

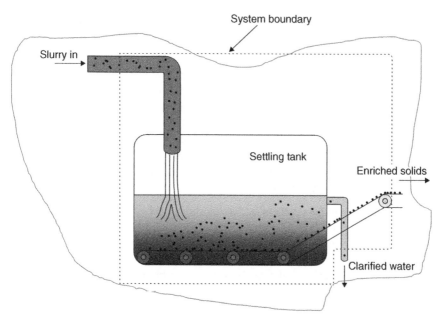

Figure 3.5 Mass balance versus time and boundary—accumulation.

2) The solids are filling the tank and not being removed.
3) An unanticipated chemical reaction generating a gas, which is leaving the system through a vent pipe or building up pressure within the tank.

You might think of other possibilities as well. These kinds of consequences, none of them desirable, are the key reasons that an online material balance in a manufacturing process is critical and usually incorporated into its process control and measurement systems.

Energy Balances

The concepts here are the same, but the factor of heat being generated or absorbed during a chemical reaction must be taken into account. We can express this type of balance as

Energy in = energy out + energy generated + energy stored

Energy generated could be negative if the reaction is endothermic (heat absorbing, more later).

In the previous example of the salt water evaporator, the mechanism of boiling the water away from the solution is to introduce steam, allowing it to "condense" and giving its energy value to the salt solution in sufficient quantities to boil the solution. If the amount of steam we introduced was less than this amount, the solution would not boil, but would merely get warmer. When we are discussing the energy balance aspects of this diagram, there are some more things that must be considered:

1) The energy content of the steam will depend on its pressure.
2) How much heat the solution can absorb will depend on its heat capacity, which in turn, will depend on its salt concentration.
3) The boiling temperature will be affected by the pressure in the evaporator as well as the liquid "head" in the evaporator.
4) Energy can be lost to the surroundings. This will be affected by what type and the amount of insulation and the temperature difference between the vessel and its surroundings.

If we have a chemical reaction occurring during the process, it could be exothermic (heat generating) or heat consuming (endothermic). The values of these reaction energy releases would be taken into account, along with the temperatures, flows, and heat capacities in calculating an online energy balance. If these numbers do not "close" or balance, there are a number of possibilities:

1) Flows are not as expected, not only in an absolute sense but also in a ratio or stoichiometric sense.
2) Temperature measurements are not correct.

3) External energy (heating or cooling) input is not what is planned.
4) If the flows are incorrect, then what may happen to the heat generation or release that is expected?
5) Lack of complete understanding of the physical properties of the materials, such as heat capacity or thermal conductivity.
6) Lack of understanding of the reliability of supporting utility systems such as refrigeration and cooling water.
7) Lack of consideration for physical mixing energy input into a system from internal process equipment such as agitators.

Momentum Balances

The energy represented by fluids and gases in motion (or for that matter, *anything* in motion) must also be conserved and balanced. We use the term momentum to describe this property. For example, if a process valve were to be suddenly shut, when it is normally open with a liquid or gas flowing through it, the energy represented by the flowing liquid will be dissipated in some way. The pipe may vibrate or burst, for example. Consider the case of a simple pipe diameter change, with a fluid (gas or liquid) moving, as seen in Figure 3.6.

Since the mass of fluid flow does not change when the pipe diameter changes, the velocity of the fluid must increase to maintain the same flow rate. In effect, the product of mass × velocity remains the same, and the momentum balance is "closed." The larger the ratio of pipe diameter change, the greater the increase in velocity will be. The opposite would be true if we expanded the pipe diameter—the velocity would slow in proportion to the diameter change.

If we were pumping liquid at a certain flow rate into a system, there is a certain amount of energy associated with this flow. There will be pressure drop in the pipeline and across valves, which when subtracted from the initial input "momentum," should equal the outlet pressure of the stream. If it does not, what might be happening?

1) The flow measurements are not correct.
2) The pressure measurements are not correct.
3) The pipe is leaking!

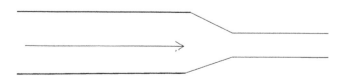

Figure 3.6 Reducing pipe size.

The details of some of these concepts, that is, fluid flow, heat transfer, and evaporators, will be discussed in more detail later, but these basic concepts regarding balances are essential to any chemical engineering analysis, and courses related to these concepts are usually the first courses taken in chemical engineering curricula.

In all cases, the laws of conservation of matter, energy, and momentum must be followed. This is true no matter what aspect or unit operation of chemical engineering we are discussing.

Momentum (fluid energy) and energy balances must also be defined in terms of time and scope as we discussed in detail for mass balance concepts.

Summary

It is essential that complete balances on mass, momentum, and energy for all equipment and processes be known, understood, and continuously calculated and updated. The effects of physical property changes must be considered. They must be tied into appropriate alarm and safety systems. It is also critical that operating personnel receive sufficient education about the processes to be able to observe and respond to situations where balances do not "close."

Coffee Brewing: Balances

How does the concept of balances apply here? Have you ever overfilled a water reservoir in a coffee brewer? Overfilled a coffee cup? What goes in must come out—eventually. Most coffee brewers have hot plates on which a carafe sits. Where does this energy go? For one thing it supplies enough heat to "balance" the heat loss from the carafe to the surrounding environment. The amount of energy is usually more than that, causing the water in the coffee solution to evaporate. If we were able to collect the evaporated water and measure the heat lost by the carafe, we would find it equal to the energy supplied to the hot plate. Have you ever put an extra filter in a machine to make sure that there are no coffee grounds dropping into the cup? You probably remember a few times when the water overflowed before it reached the carafe. The momentum of the water dripping into the filter was constant, and when the filter supplied too much back pressure, the momentum was maintained by the overflowing liquid.

Discussion Questions

1   Are all the mass and energy balances around each of your unit operations measured online and understood? What physical properties would affect these calculations, and are they known or measured? Or just estimated? How "off" do these properties need to be to have a problem? What kind of problem?

2   Are all aspects of fluid and gas energy accounted for? What happens if flow rates increase or decrease? What physical property data changes might affect?

3   Are inventory tanks properly monitored? Are flows in and out measured and balance continuously made? What is the response if the balance does not close?

4   How does your process control and instrumentation calculate mass, energy, and momentum balances? How are discrepancies reported? Handled? Alarmed?

5   When your plant makes changes in your process, how are these concepts of balances used? How could they be used? Are they sufficiently included in other review processes?

6   If there was a major discrepancy in a balance calculation (either manual or via instrumentation), what are possible environmental consequences? How are they monitored?

Review Questions (Answers in Appendix with Explanations)

1   The concept of balances in chemical engineering means that:
    A __Mass is conserved
    B __Energy is conserved
    C __Fluid momentum is conserved
    D __All of the above

2   If a mass balance around a tank or vessel does not "close" and *instrumentation readings are accurate*, then a possible cause is:
    A __A reactor or tank is leaking
    B __A valve or pump setting for material leaving the tank is incorrect

    C __A valve or pump setting for material entering the tank is incorrect

    D __Any of the above

3 If an energy balance around a reaction vessel shows more energy being formed or released than should be (*and the instrumentation readings are correct*), a possible cause is:

    A __Physical properties of the materials have changed

    B __A chemical reaction (and its associated heat effects) is occurring that has not been accounted for

    C __Insulation has been added on the night shift when no one was looking

    D __A buildup of material is occurring

4 If pressure in a pipeline has suddenly dropped, it may be because:

    A __A valve has been shut not allowing fluid to leave

    B __A valve has been opened, allowing fluid to leave

    C __It has calmed down

    D __A downstream process has suddenly decided it would like what is in the pipe

5 Ensuring accurate measurements of pressure, flow, and mass flows is critical to insure:

    A __We know what to charge the customer for the product made that day

    B __We know when to order replacement parts

    C __We know how to check bills from suppliers

    D __Knowledge of unexpected changes in process conditions

Additional Resources

Hatfield, A. (2008) "Analyzing Equilibrium When Non-condensables Are Present" *Chemical Engineering Progress*, 4, pp. 42–50.

Ku, Y. and Hung, S. (2014) "Manage Raw Material Supply Risks" *Chemical Engineering Progress*, 9, pp. 28–35.

Nolen, S. (2016) "Leveraging Energy Management for Water Conservation" *Chemical Engineering Progress*, 4, pp. 41–47.

Richardson, K. (2016) "Predicting High Temperature Hydrogen Attack" *Chemical Engineering Progress*, 1, p. 25.

Theising, T. (2016) "Preparing for a Successful Energy Assessment" *Chemical Engineering Progress*, 4, pp. 44–49.

4

Stoichiometry, Thermodynamics, Kinetics, Equilibrium, and Reaction Engineering

We will now cover several topics that affect some of the practical aspects of scaling up a chemical reaction and the integration of chemistry and chemical engineering.

Stoichiometry and Thermodynamics

First, there is the general topic of *stoichiometry*. This is a word, derived from Greek, which describes the ratio and amounts of chemicals that react with each other in a chemical reaction. Here we must introduce some basic chemistry concepts. Every chemical has a different molecular structure, size, and weight as determined by its molecular content. In the early 1900s, a brilliant Russian chemist, Mendeleev, was able to organize the known chemical elements according to their atomic number, atomic weight, as well as by the nature of their chemical activity. By adding the number of protons in a molecule and the corresponding number of electrons that balance the charge, its atomic number is determined. For example, carbon, C in the table, has an atomic number of 6 and an atomic weight of 12. The table is grouped into types of elements with similar chemical behavior. For example, the "active" metals, such as lithium (Li), sodium (Na), and potassium (K), are in the same column. We see that the halogens fluorine (F), chlorine (Cl), and bromine (Br) are also grouped together. We see the "inert" gases such as helium (He), neon (Ne), argon (Ar), and krypton (Kr) also grouped together. Some gaseous molecules such as nitrogen, oxygen, chlorine, and bromine exist, at normal conditions, as *diatomic* molecules (N_2, O_2, Cl_2, and Br_2, respectively). In these cases the *molecular* weight will be twice the atomic weight. For example, the atomic weight of nitrogen is 2×14 (its atomic weight) or 28. Chlorine's would be 71 or 2×35.5. These distinctions are important as chemicals react according to their molecular weight, not their atomic weight. In some cases these are the same, but in these cases, they are not. A *mole* is the amount of a given chemical equal to its molecular weight,

Chemical Engineering for Non-Chemical Engineers, First Edition. Jack Hipple.
© 2017 American Institute of Chemical Engineers, Inc. Published 2017 by John Wiley & Sons, Inc.

expressed in any units that are consistent. For example, one gram mole of diatomic chlorine (Cl_2) is 71 g, one gram mole of diatomic hydrogen (H_2) is 2 g, and one gram mole of diatomic nitrogen (N_2) is 28 g. We could also express these as pound moles if we were working in the English system. A mole is a critical concept to understand as chemicals react as moles, not as weights. Weight is a secondary function, not the primary one. It is possible to produce diatomic molecules monoatomically but only under extreme and unusual conditions (Table 4.1).

Another very basic concept is that, when writing a chemical reaction equation, the molecular weights on both sides of the equation must "balance." Consider the example of burning carbon in air as described by this equation:

$$C + O_2 \rightarrow CO_2$$

We have one carbon on each side of the equation and two oxygen molecules on both sides, so the equation "balances." If we take into account the atomic or molecular weights of the molecules, this also balances:

$$C + O_2 \rightarrow CO_2$$
$$(12) + (32) \rightarrow (44)$$

A balanced equation does not mean that the chemical reaction we have described will happen. For example, if you expose a piece of charcoal (virtually all carbon, C) to air (21% oxygen, O_2), does it burn and produce carbon dioxide? No—it needs a "spark" or initiation of some sort.

As another example, many homes are heated by natural gas (primarily methane or CH_4) via a combustion process. We can show a simplified version of this reaction as

$$CH_4 + 2O_2 \rightarrow CO_2 + 2H_2O$$

This reaction is exothermic, or "heat generating," heating the air going through the furnace that then flows through the ductwork and heats the house. This chemical equation is balanced in the sense that it has the same number of carbon, hydrogen, and oxygen atoms on both sides, and so it will also balance from a weight standpoint. But does natural gas burn itself? No, your furnace or your gas-fired hot water heater, using natural gas as fuel, has a pilot light to initiate this reaction, and once done, the reaction continues. We will discuss the concept of initiating a reaction in the next section.

Another important point is that just because someone has written a balanced equation does not mean that it will *ever* occur. It just says that if it occurs, this is a possible outcome. For example, we know it is possible to burn natural gas and also produce carbon monoxide (an odorless, poisonous gas) by also burning natural gas:

$$2CH_4 + 3O_2 \rightarrow 2CO + 4H_2O$$

Table 4.1 Periodic table of elements.

Group →	1	2	3	4	5	6	7	8	9	10	11	12	13	14	15	16	17	18
↓ Period																		
1	1 H																	2 He
2	3 Li	4 Be											5 B	6 C	7 N	8 O	9 F	10 Ne
3	11 Na	12 Mg											13 Al	14 Si	15 P	16 S	17 Cl	18 Ar
4	19 K	20 Ca	21 Sc	22 Ti	23 V	24 Cr	25 Mn	26 Fe	27 Co	28 Ni	29 Cu	30 Zn	31 Ga	32 Ge	33 As	34 Se	35 Br	36 Kr
5	37 Rb	38 Sr	39 Y	40 Zr	41 Nb	42 Mo	43 Tc	44 Ru	45 Rh	46 Pd	47 Ag	48 Cd	49 In	50 Sn	51 Sb	52 Te	53 I	54 Xe
6	55 Cs	56 Ba	71 Lu	72 Hf	73 Ta	74 W	75 Re	76 Os	77 Ir	78 Pt	79 Au	80 Hg	81 Tl	82 Pb	83 Bi	84 Po	85 At	86 Rn
7	87 Fr	88 Ra	103 Lr	104 Rf	105 Db	106 Sg	107 Bh	108 Hs	109 Mt	110 Ds	111 Rg	112 Cn	113 Uut	114 Fl	115 Uup	116 Lv	117 Uus	118 Uuo

| | | | 57
La | 58
Ce | 59
Pr | 60
Nd | 61
Pm | 62
Sm | 63
Eu | 64
Gd | 65
Tb | 66
Dy | 67
Ho | 68
Er | 69
Tm | 70
Yb | | |
| | | | 89
Ac | 90
Th | 91
Pa | 92
U | 93
Np | 94
Pu | 95
Am | 96
Cm | 97
Bk | 98
Cf | 99
Es | 100
Fm | 101
Md | 102
No | | |

Note that in the second equation, the ratio of oxygen to methane is less (3/2) versus (2/1) in the first equation. It makes sense that less oxygen present is more likely to produce more carbon monoxide. That is why any home natural gas heater is always set to use more oxygen (air) than it needs. This wastes some energy but provides a safety factor for the homeowner. Any balanced chemical equation, by itself, does not tell us what *other* chemistry may be possible.

There are many exothermic reactions that require an initiating source of energy, after which they will sustain themselves. The technical term for energy content is enthalpy, usually with the letter H or ΔH, and this will be seen in many of these types of diagrams. We can picture this as shown in Figure 4.1. The difference between the reactants and products line is the overall energy released in the reaction (when the enthalpy or energy of the system drops, this indicates energy has been released, indicating an exothermic reaction).

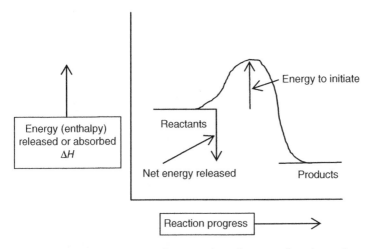

Figure 4.1 Activation energy and energy release for an exothermic reaction.

"Enthalpy" is the thermodynamic word we use to describe the energy content of a material. When enthalpy is reduced, the energy reduction is compensated for by a release of energy (what comes in must come out). The amount of this released energy is shown as the "net energy released" in Figure 4.1. The activation energy needed to initiate the reaction is also shown in the graph. This is the equivalent to the amount of energy required to be supplied by a spark plug or match in a combustion process.

If a reaction is highly exothermic and requires no initiation energy, there is a high probability that it is inherently unstable. An example would be peroxides or explosives, which are typically stored under refrigerated conditions to prevent decomposition. There would be little or no initiation required.

Let's look at the basic concepts of thermodynamics as applied to chemical reactions. As stated earlier, thermodynamics relates to the energy states and energy consumption or input of a chemical reaction system. Each element in the periodic table is assigned a net energy value of zero, reflecting its natural state. If we wish to change the form of an element, or have it react with some other element, this requires an energy change to move away from equilibrium (the natural state of an element). For example, coal (mostly carbon or "C") in the periodic table has a net energy of zero. If we burn it with air (oxygen), we produce carbon dioxide, or CO_2, what is the net energy of this reaction? Instinctively, we know that energy is released in this reaction, but how much? The "heats of formation" (ΔH_f) of most chemical compounds have been previously measured and can be found in numerous literature references or on the web. The net energy release from a reaction is the net energy contained in the molecules on the right-hand side of the chemical reaction minus that of the molecules on the left. In this simple case, since carbon and oxygen are natural elements, their thermodynamic heat value (in their natural state as gases) is zero. We can look up the heat of formation of CO_2 and find that it is a *minus* 394 kJ/mole (−394 kJ/mole). The convention in thermodynamics is that minus indicates heat release, while a plus or positive sign indicates the need for energy input. So the burning of carbon is a large net generator of energy. The fact that this number is negative is also an indication that CO_2 is a more stable state than carbon and oxygen by themselves, so if there is a way to make this reaction happen, it is preferred. However, in many cases where there is a net energy release to a more preferred state, this doesn't happen automatically. It frequently takes an input of energy to *initiate* the reaction. That's what is done with a pilot light, a spark plug, or a match. Once the reaction is initiated, there is a sustained release of energy, which will sustain the reaction as long as the supply of materials is maintained. If there is a net release of energy during a chemical reaction, we refer to it as an *exothermic* reaction. On the other hand, if the calculation of the net energy between raw materials and products is positive (meaning we need a constant input of energy to sustain the reaction), it is referred to as *endothermic*. The energy consumed in an endothermic reaction must be taken into account when calculating cost and economics. The energy output of an exothermic reaction has the potential to be reused elsewhere in a process or chemical complex. However, exothermic reactions are also inherently less safe than endothermic reactions in the sense that they produce enough energy to sustain themselves. If this energy is less than a reaction system's ability to remove heat, a runaway chemical reaction can occur. We will discuss this further in Chapter 9 when we review the concept of reactive chemicals.

An important reminder, when calculating heats of reactions, is to make sure that the heats of formation (ΔH_f) are for the compounds as they are actually going to be used in the process. For example, if water (H_2O) is a reactant or product of a reaction, it is important to specify whether the water is in solid, liquid, or vapor form as the heats of formation of water and steam (water vapor) are significantly different due to the energy required (44 kJ/mole) to boil liquid water and turn it into a gas. The freezing or melting of water also has an energy change associated with it.

A general comment about units at this point in time is necessary. We stated the energy release of the carbon dioxide reaction as a −394 kJ/mole. We could have stated this in terms of BTUs, calories, kcals, or any other number of unit forms. In today's global world with competing unit systems (British, metric, SI), it is *imperative* that any technical value be consistent, in terms of units used, with other technical values in the same system. There is no right or wrong about any choice, but it must be consistent or serious consequences in terms of chemical reactivity and process control can result. This is especially important within multinational corporations whose calculations, drawings, and memos may be generated in many different countries. This warning and caution applies to all future subjects to be discussed, as every chemical engineering unit operation will have activities expressed in terms of scientific units that can be expressed in multiple ways. Serious safety and design consequences can result if one part of a team within a multinational organization is thinking and working in English units and the other working and communicating in metric or SI units.

Kinetics, Equilibrium, and Reaction Engineering

Let's first review the concept of equilibrium. Any chemical reaction, even with a large negative energy output, does not necessarily go to completion, meaning that all of the reactants convert to all the desired products. When a chemical reaction system reaches equilibrium, it may look like the system is stagnant, but in actuality the forward and backward reactions are going on at the same rate. The point at which this happens for any chemical reaction system will be affected by temperature as well as the ratios of reactants, and for gases, the pressure as well. For an endothermic (heat consuming) reaction, the equilibrium will almost always move further to the right (i.e., more conversion) as the temperature of the reaction is increased. There also may be a point, if the temperature is increased, that the products may begin to decompose into other products, usually undesirable. For exothermic (heat releasing) reactions, there will be a temperature reached at which the equilibrium shifts back in favor of the reactants. The effects of pressure, when gases are involved in the reaction, will be discussed later.

Another pair of terms used in describing chemical reactions is *reversible* and *irreversible*. A reversible reaction would be one that is at equilibrium but can be reversed through a change in process conditions such as temperature or pressure. A gas–gas reaction might be of this type. An irreversible reaction describes a reaction that cannot be reversed. This would normally be the case, where a liquid–liquid reaction produces a gas, which escapes and cannot be recaptured or recycled, or when such a reaction produces a solid, which precipitates out of solution and cannot be redissolved without significant change in process conditions.

A French chemist Le Chatelier, in the 18th century, made a key observation about chemical reactions and that was that they reacted and changed to relieve "stress." For example, if we raised the amount of one of the reactants on the left-hand side of a chemical reaction equation, and all else being equal, the system would react by trying to minimize the effect of this change by moving the reaction further to the right (product) side. The same would be true if we added additional product to the right-hand side of the reaction equation; the reaction would shift back to the left in an attempt to maintain equilibrium. This concept is very useful in a practical chemical engineering sense in that it provides options to push a reaction further to completion to improve conversions and yields as well as giving us a fundamental understanding of how a reaction system will qualitatively respond to changes in stoichiometry, pressure, and temperature.

An industrial example illustrating all of these points is the manufacture of sulfuric acid (H_2SO_4). The first step in the currently used contact process is the combustion of sulfur (usually produced from mining an underground deposit or possibly recovered from another process). The equation for this exothermic reaction (generating 297 kJ/mole—note the minus sign in front of the amount of energy meaning energy is released; an exothermic reaction) is

$$S + O_2 \rightarrow SO_2 \quad \Delta H = -297 \text{ kJ/mole}$$

As we discussed earlier, this is a highly exothermic reaction but still needs a "spark" to ignite the sulfur. Sulfur does not start burning by itself. However, once ignited, the sulfur will burn continuously (producing SO_2 and releasing energy) until either the air or sulfur is withdrawn or consumed.

The second step in this process is to convert the SO_2 into SO_3 (sulfur trioxide) via the following reaction, also exothermic (but less so than the first reaction):

$$SO_2 + \tfrac{1}{2}O_2 \rightarrow SO_3 \quad \Delta H = -197 \text{ kJ / mole}$$

This reaction requires a catalyst, typically based on vanadium pentoxide (V_2O_5), and the reaction is a *heterogeneous* (two different phases) catalytic reaction in that the sulfur dioxide gas is passed over pellets of catalyst, similar to what happens in your car exhaust catalytic converter. In this reaction, as in many other exothermic reactions, the yield to SO_3 *decreases* as the temperature increases. In order to achieve complete conversion of SO_2 to SO_3, the temperature must be reduced. In the actual process this is done in stages with cooling water to eventually achieve nearly 100% conversion. This, unfortunately, reduces the amount of possible reuse of the heat generated during this reaction. This is a classic contradiction in many exothermic chemical reactions.

Finally, the sulfur trioxide is reacted and absorbed into water, producing sulfuric acid (H_2SO_4), in another exothermic reaction:

$$SO_3 + HO_2 \rightarrow H_2SO_3 \quad \Delta H = -130 \text{ kJ/mole}$$

In practice, the SO_3 is absorbed (more about the unit operation of absorption later) into already produced concentrated sulfuric acid, producing what is known in this industry as "oleum," or fuming 100% sulfuric acid, which is then later diluted to the concentration desired. Any unabsorbed/reacted SO_3 must be recycled into the previous reaction step.

A simple block flow sheet of how all of these reactions are linked together is shown in Figure 4.2.

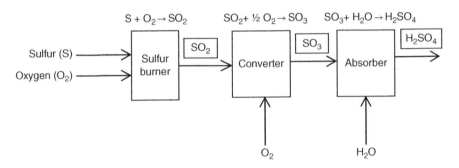

Figure 4.2 Manufacture of sulfuric acid via contact process.

The converter unit operation is far more complex than the other steps due to the equilibrium limitations discussed earlier. In order to achieve conversions above 97%, the reaction gases are cooled in a series of stages to push the equilibrium to the right. This wastes some of the exothermic reaction energy but is the only way to achieve a 100% conversion of SO_2 to SO_3. A detailed diagram of this last process stage is shown in Figure 4.3. At the entrance to the converter step, the incoming SO_2 is preheated with the outlet gas from the final exit gas from the converter step to raise the inlet gas temperature to the final step to 730 K (or about 460°C). As the final reaction begins to occur, further heat is released, but now the equilibrium begins to limit the conversion. To overcome this, the reaction gases are cooled to shift the reaction equilibrium to the right. This is done in a series of stages, and the process design attempts to recover as much of the reaction-generated heat as possible. This is a traditional chemical engineering design compromise between kinetics and thermodynamics.

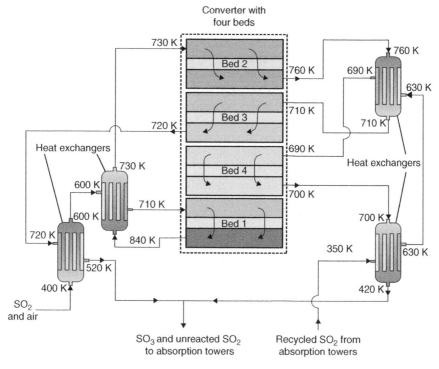

Figure 4.3 Contact sulfuric acid process converter step. Source: Reproduced with permission of Centre for Industry Education Collaboration.

All of these topics we have discussed can be grouped under the general heading of chemical and reaction thermodynamics. They help us understand how much energy is released or consumed during a reaction, which reactions are favored, and how to enhance a reaction system to produce the most desired products and minimize raw materials and energy consumption. Most of these properties are measured in laboratory experiments or obtained from reliable literature resources and are indispensable to chemical engineers in optimizing a full-scale chemical process.

Physical Properties Affecting Energy Aspects of a Reaction System

Since energy is absorbed or released in a reaction (as well as in other chemical process operations such as pumping and agitation), we must be able to calculate the effects of these energy changes on the system. Temperature change in a system is determined primarily by its inherent ability to absorb heat. The physical property that determines a material's ability to absorb and contain heat is its *heat capacity*. This physical property has the units of energy/unit mass/unit time, frequently designated as C_p and expressed as BTU/#/°F or cal/g/°C. Water is used as a standard and has a heat capacity of 1 BTU/#/°F at room temperature. This property does change slightly with temperature but not to a significant degree. A material with a higher heat capacity will be able to absorb more heat without raising temperature, and conversely, one with a lower heat capacity will see a greater rise in temperature for the same thermal input. This is important when we are thinking about exothermic reactions and how to contain the heat that is generated.

Here are some examples of heat capacities of some common substances at 25°C.

Table 4.2 Heat capacities of common materials.

Water	1.0
Ethyl alcohol	0.6
Graphite	2.1
Oxygen	7.0
Nitrogen	7.0
Methane	8.4
Salt (NaCl)	12.2

Source: Average of publicly available information.

In general, the heat capacities of solids are greater than liquids, which in turn are greater, in general, than gases. In comparing ethanol and water in the previous list, we can say that water can absorb about 2/3 more energy per unit mass than ethanol. If we used water versus ethanol, say, in a heat exchanger, it would take roughly 1.5 times as much ethanol to contain the same amount of heat rise. The heat capacity will have a direct impact on how much energy a storage or reaction system can absorb or could release.

Two equally important properties in the sense of a material's ability to absorb heat are its heat of fusion/melting (ΔH_f), which is a measure of how much energy change is seen when a solid melts or freezes, and its heat of vaporization (ΔH_v), the amount of energy it takes to boil a material or, conversely, how much energy it releases when it condenses. In a chemical process design, this property can have a major impact on the ability of a system to maintain temperature at a constant point, especially if the boiling of a liquid is being used as a temperature control mechanism.

Another material property is its *thermal conductivity*, *k*. This property is usually expressed in units of energy/unit time/ΔT, or BTU/h/°F (cal/s/°C). As opposed to heat capacity, which measures a material's ability to absorb heat, thermal conductivity measures how rapidly heat moves *through* a material at constant temperature. The most common place we think about this property is when we are evaluating how much insulation to install in a house to prevent heat loss, or in the chemical industry, how much insulation is installed around hot pipelines, reactor vessels, or buildings. In a chemical process, *k* will affect the rate at which heating or cooling can be supplied to a reaction vessel, as well as the capital cost of insulating equipment or piping and the resulting savings. Examples of thermal conductivity values are shown in Table 4.3.

Table 4.3 Thermal conductivities of materials.

Hexane	0.08 BTU/h/°F
Water	0.34–0.38 BTU/h/°F
Sodium metal	45–50 BTU/h/°F
Hydrogen	0.10–0.12 BTU/h/°F
Methane	0.18–0.22 BTU/h/°F
Air	0.014–0.018 BTU/h/°F
Argon	0.01 BTU/h/°F
Carbon dioxide	0.025 BTU/h/°F

Source: Average of Public Source Information.

The extraordinarily high thermal conductivity of sodium metal is the primary reason it is used as a heat sink in commercial nuclear power plant reactors in the event of an emergency situation. The 40%+ lower thermal conductivity of argon versus air is the reason that argon-filled window panes are used in extreme northern climates to minimize heat loss in the winter.

Kinetics and Rates of Reaction

We have discussed aspects of a chemical reaction in terms of its net energy release and its ultimate equilibrium. But how fast does a reaction reach its endpoint or equilibrium? That is determined by a reaction's *kinetic rate constant*, usually designated by *k*. Its units are typically moles/s or moles/h for slow reactions. An example of a fast reaction rate would be the combustion of carbon (in the form of wood) in a forest fire versus a slow reaction rate as exhibited by the oxidation of iron to produce rust:

$$2Fe + 3O_2 \rightarrow 2Fe_2O_3$$

Both are harmful and are oxidations, but one occurs at a much faster rate than the other.

Very slow reaction rates, involving very slow degradation or decomposition of materials in storage, may not be readily obvious. It is always important to be alert to signs of such reactions, which may take the physical form, for example, of bulging drums or corrosion deposits external to a pipeline. It is also possible to ignore expiration dates on raw materials in storage or products in inventory for shipping.

If there are multiple reactions occurring, each will have its own kinetic rate constant, and if the chemistries occurring in the reaction system have overlap in their raw materials and products, then the rate of each reaction will affect that of others. For example, consider the chemistry of nitrogen oxides. The first step in the manufacture of nitric acid is the oxidation of ammonia:

$$4NH_3 + 5O_2 \rightarrow 4NO + 6H_2O$$

The NO produced is further oxidized to nitrogen dioxide (NO_2):

$$2NO + O_2 \rightarrow 2NO_2$$

The nitrogen dioxide (NO_2) is then reacted and absorbed into water to produce nitric acid (HNO_3) via this reaction:

$$3NO_2 + H_2O \rightarrow 2HNO_3 + NO$$

Since this last reaction produces NO in addition to the desired product, we need to find a way to recycle it back into the second reaction, which uses NO as a feedstock.

Kinetic rate constants tend to be *logarithmic* with temperature, that is, they increase exponentially with temperature. If we were to plot a kinetic rate constant as a function of temperature for almost any reaction, we would see a curve that looks similar to Figure 4.4.

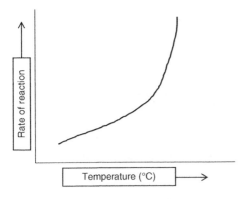

Figure 4.4 Rates of reactions accelerating as temperature increases.

If we were to plot this as a semilogarithmic plot, with the log of the kinetic rate constant plotted against the inverse of the absolute temperature, we would see a graph similar to that in Figure 4.5.

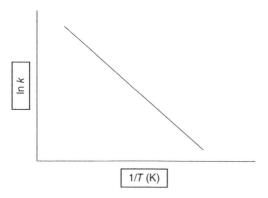

Figure 4.5 Rates of reaction versus temperature (semilog plot).

The slope of this line represents the activation energy (how much energy does it take to initiate the reaction?) for the reaction. The greater the slope, the higher the activation energy, meaning that it will take a greater amount of energy to initiate the reaction. Remember though, for an exothermic reaction, once it has been initiated, the reaction will continue without further energy input. For many systems, the slope of this line represents a doubling of reaction rates every 10°C. The greater the slope, the more sensitive the reaction rate is to temperature. The lower the slope, the less sensitivity and the less activation energy required to initiate the reaction.

When the chemistry is simple, the relationship between the stoichiometry and rate of reaction clear, we may see descriptions of reactions as being "zero," "first," or "second" order, referring to the response of reaction rate to the concentration of reactants, as opposed to temperature, which is still the dominant factor.

For example, if a reaction had involved A + B to produce a single product C and the reaction rate, k, was proportional to the concentrations of both A and B, respectively, we would say the reaction was first order with respect to A, first order with respect to B, and second order overall. If the reaction was proportional to the concentration of A but proportional to the concentration of B to the second power (B^2), we would say that the reaction rate was first order with respect to A, second order with respect to B, and third order overall. There are some reactions that are zero order, meaning their rates respond only to temperature and not the concentrations of reactants. An example of this is the decomposition of carbonic acid (H_2CO_3) to carbon dioxide and water (this is the loss of carbon dioxide in a soda pop, which loses its "fizz" when left out of the refrigerator too long):

$$H_2CO_3 \rightarrow CO_2 + H_2O$$

For these basic reaction types, here is a list of plots, which would be a straight line:

Zero order: concentration versus time (t)
First order: concentration versus $1/t$
Second order: $1/$concentration versus t

In qualitative graphical format, the approximate change in concentration of component A, decomposing according to these various rate laws, would look as shown in Figure 4.6.

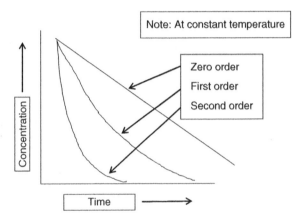

Figure 4.6 Concentration of reactants versus time as a function of reaction order.

The sharper the drop-off in initial reactant concentration, the higher order is the reaction and the higher is the reaction rate constant. These two factors combine into a reaction rate expression equivalent to

$$r = kA^aB^b$$

The sum of a and b is the order of the reaction. Remember that these exponents could be any kind of mathematical figure including square roots. None of these general descriptions describe in any way the detailed molecular interactions that are going on, only an empirical representation of the end result.

There is some standard terminology used in the chemical industry to describe some of the concepts we have discussed. Unfortunately, some of these terms are confused with each other and often used interchangeably. Here is the list suggested for your use:

Reaction Conversion. How much of the primary raw material fed into the reactor reacted to something (not necessarily the product of desire)

Reaction Yield. How much of the reacted raw material was converted to the product desired

Selectivity. Ratio of the desired product produced divided by all the products produced

High conversion, yield, and selectivity are always what we are striving for in a chemical reaction system unless we desire to produce a distribution of products.

Other terms seen in the description of reaction systems include "reversible" and "irreversible." In the case of gas-to-gas reactions, a simple change of temperature or pressure can reverse a reaction since nothing is condensed or removed. An irreversible reaction would be one in which one of the products changes phase (vaporizes, precipitates, condenses).

In summary, thermodynamics determines the *possibility of a reaction* and under what conditions, and kinetics determine the *rate at which it occurs*.

Catalysts

Catalysts are materials that do one of several things:

1) Allow a chemical reaction to occur under less severe conditions than normal or promote a reaction that does not occur at all at reasonable conditions. For example, the unburned hydrocarbons in your car's gasoline engine need to be reduced to meet current EPA emission guidelines. These unburned hydrocarbons could have been eliminated through the use of some kind of an afterburner, but that would have added a great deal of cost and complexity to the car's engine and exhaust systems. The catalytic converter currently installed in your car provides an "activating" catalytic surface, containing platinum, upon which the hydrocarbons and oxygen can react at the temperatures present in the exhaust stream. This unique catalytic surface provides a pathway for the unburned hydrocarbon to react with oxygen in the air to produce carbon dioxide and water. The net result is the same as any other hydrocarbon oxidation process to carbon dioxide and water but at a lower temperature.

Another example is the reaction of nitrogen (N_2) and hydrogen (H_2) to produce ammonia:

$$N_2 + 3H_2 \rightarrow 2NH_3$$

This reaction does not proceed to any measurable degree under room temperature and atmospheric pressure, but with an iron-based catalyst and high pressure (>1000 psig), the reaction yields a 15–20% conversion to ammonia. The discovery of this catalyst, which indirectly feeds a large share of the world's population, was awarded the Nobel Prize in 1919. In these situations, what the catalyst is doing is lowering the "energy of activation" (or, more accurately, providing a lower energy pathway) to allow the chemistry to proceed (refer back to Figure 4.1).

2) Change the rate of a reaction. A catalyst can change the rate of a reaction significantly without necessarily changing the outcome of the reaction. If we mix hydrogen (H_2) and bromine (Br_2) together at room temperature, virtually nothing happens. If, however, we pass these gases over a platinum metal catalyst at 300 K, the rate is enhanced many orders of magnitude. Note that the final product of the reaction does not change, only the rate.

3) Change the selectivity reactor conditions to produce one product in favor of another. An example here might be the reaction of ethylene oxide with ammonia to produce a class of materials we call ethanolamines. These compounds are extremely effective in absorbing carbon dioxide (CO_2) and hydrogen sulfide (H_2S) from sour natural gas to produce "sweet" gas, which can then be put into the natural gas pipeline distribution system. The chemistry used to manufacture these amines is shown in Figure 4.7.

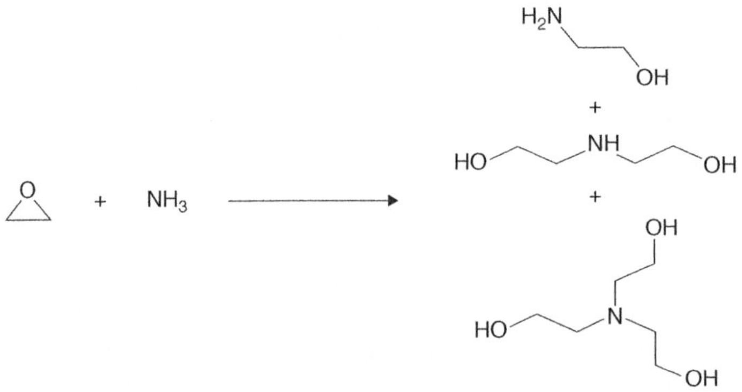

Figure 4.7 Reactions of ammonia with ethylene oxide to produce ethanol amines.

Each of these reactions has a different reaction rate constant and using a catalyst, which enhances one reaction over another, can allow a change in the product distribution according to customer and business needs.

Catalysts, in general, do not change the outcome of the reaction; they merely change the conditions under which the process operates by providing a different pathway for the chemistry to occur. They *can* change the favorability of one reaction over another when multiple reactions are possible.

There are two general classes of catalysts, homogeneous and heterogeneous. *Homogeneous* refers to a catalyst that is dissolved in the reaction solution. The most common of these are organometallic compounds. They are used in the manufacture of materials such as acetic acid and acetic anhydride and in some polymerization processes. They are typically very sensitive, and are deactivated, by water.

A *heterogeneous* catalyst is one that is in a separate phase from the primary reactants. The catalytic converter in our cars is such a catalyst since exhaust gases flow over a platinum metal coated on a ceramic substrate. Many gas phase polymerization processes use such a catalyst. Such a catalyst could also be a solid in suspension in a liquid–liquid or gas–liquid reaction.

Catalysts do not sustain their performance forever and can be poisoned by certain contaminants. Poisoning refers to a chemical reaction that renders the catalyst nonactive or coats its surface in a way that makes the catalytic surface not accessible to the reactants. There can also be physical degradation of solid catalysts at grain boundaries within their structures. The platinum catalyst in the automobile that converts unburned hydrocarbons into carbon dioxide and water is poisoned by lead (Pb), which is one of the reasons we now use unleaded gasoline in our cars. One of the first major operations in a gasoline refinery is the catalytic cracking of its primary feedstock into lighter, more valuable components. This catalyst is poisoned, though not irreversibly, by carbon and needs to be shut down and regenerated, by burning off the carbon, on a regular basis. Sulfur is also a well-known poison for many hydrocarbon reactions so the sulfur in high-sulfur oil or the hydrogen sulfide in natural gas must be removed prior to downstream processing.

Physical aspects of a chemical reaction can significantly affect the equilibrium of a reaction. These types of reactions are normally *irreversible*:

1) Two (or more) liquids or a gas–liquid reaction that produces a solid precipitating out of solution
2) Oxidation reactions, that is, combustion, such as the burning of a fuel

When there is no change in phase, it is probable that the reversibility concepts we have reviewed need to be considered.

Summary

The fundamentals of chemistry, kinetics, and physical properties are key aspects that determine many of the design parameters of a chemical reaction system. It is easy to overlook the impact of changes in properties, especially with respect to how they change with temperature. This especially applies to reaction rates and gas volumes.

Coffee Brewing: Is Kinetics Relevant?

Let's now look at the coffee brewing process from a materials, reaction, and chemical kinetics point of view. As we have discussed, reaction rates are, in general, a strong function of temperature. We all recognize that "fresh" coffee tastes better than "old" or "stale" coffee, possibly sitting on a hot plate for a long time. Why is that? Coffee, as well as any other food product, is a composite of chemicals. We drink coffee both because of its taste and its ability to keep us awake. How does it do this? It contains caffeine, a chemical, which for most people, is a stimulant. But how does the coffee taste after sitting on a hot plate for a few hours? Is it the same? Unless someone is very desperate for caffeine, the answer is "no" and new pot will be brewed. What is going on is a chemical degradation of some of the ingredients of the coffee into a class of chemical compounds known as aldehydes and ketones. The longer the coffee sits on the hot plate, the greater the extent of this chemical degradation. Since most people have no interest in cold coffee, the temperature of the hot plate is not reduced and the chemical degradation continues, and the degradation (a chemical reaction!) increases rapidly with temperature. As the coffee sitting on the hot plate evaporates, the concentrations of the materials that are degrading increases, raising the chemical degradation rate due to concentration increase.

What are some approaches that you have seen used to minimize this issue? There are coffee carafes that are basically heavily insulated vacuum containers that do not sit on a hot plate. Is degradation still occurring? Yes, but the concentration of the degrading solution is minimized due to the lack of evaporation and degradations (chemical reactions) are proportional to concentration. We now have individual coffee brewing systems that only brew one cup at a time with nothing left to degrade in taste or quality. Since these machines brew coffee so rapidly, the inconvenience of having to wait less than a minute for a fresh cup of coffee is tolerable versus the alternative of being able to pour a new cup instantaneously, but that cup tasting far worse.

When you buy coffee, what kind (not brand) of coffee do you buy? Instant? Freeze dried? Ground? Beans? Ground in the store? Stored at home and then ground? Where stored? Freezer? Refrigerator? Normal cabinet? What is the difference, from a chemical kinetics standpoint, between them? Why do they taste differently?

Do we envision coffee brewing in our homes as a reversible or irreversible process? Though the primary step of brewing is not a chemical reaction (it is a leaching process, to be discussed later), do we think of this process as reversible? In other words, do we think that if the coffee does not taste the way we prefer it, we can easily reverse the process and start all over again? We would have to evaporate the coffee, dry the grounds, etc. essentially producing a recycled version of instant coffee!

(Continued)

(Continued)

When we decide to brew a cup of coffee, we have a "recipe" in mind (i.e., "stoichiometry"). We may like weak- or strong-tasting coffee. Unsuspected impurities in any of the raw materials (coffee, water, additives) can cause health problems. We may care about an added flavor or caffeine levels. In any of our home brewing systems, we have a choice of coffee serving size, and given a fixed amount of coffee dumped into a filter or in a "pod," this is decided by how much water is used and at what temperature it is delivered. All of these individual choices will affect the outcome of the "process" we are running (brewing coffee), just as in a real chemical process. The water that is used in the brewing process can come from any number of sources, including tap water (whose quality and impurities vary all over the world), "spring" water purchased in a store (what spring?), distilled water, or water that has been run through an attachment to a faucet to remove "impurities." The variation in these raw materials will affect the final product no differently than changing raw materials entering a chemical process. Just as a chemical plant's customers desire a consistent product, according to the specifications agreed upon in a contract, produced from whatever raw materials the supplier uses, the coffee drinker desires the same taste in their final cup they drink.

So we have a compromise, not uncommon in reaction systems, between our desire to have product at a high temperature (hot coffee) and parallel reactions that may degrade the product if left at a high temperature for an extended time. More as we go along.

Discussion Questions

1 Is the detailed stoichiometry of your process known? Are there intermediate steps in the reactions, the study and understanding of which could improve process efficiency or product quality? Does everyone involved in the operation of a process plant have a basic understanding of the process chemistry?

2 Are the consequences of poor stoichiometry control well understood? From a quality standpoint? From a reactive chemical standpoint? From a process control standpoint?

3 Are the kinetics and rate constants of the reactions being run understood? If not, what is the basis for deciding on a reaction "recipe" or sequence? Have deviations from this "recipe" ever given signs of a problem (quality, safety)?

4 Thinking about your current commercial processes—which ones experienced start up problem due to a lack of understanding of process chemistry? Is improving this understanding part of your organization's R&D program? If not, why not?

5   Are the thermodynamics of all your process steps understood? Which reactions are exothermic? Endothermic? If exothermic, is the crossover point, where the reaction rate generates more heat than is possible to remove, known? What prevention steps have been taken? How many layers of protection are needed? How was this decided?

6   Are all the basic physical properties of the materials you are using and generating (heat capacity, density, thermal conductivity, viscosity) known? Is their sensitivity to temperature and pressure known and applied in how the process is run? How could changes in raw materials fed into a reaction change these values during a reaction or for the products produced?

7   Are the equilibrium constants for the reactions known? Is it well understood how temperature and pressure changes can affect these values? Are these effects important? If so, how is the information used? If not, why not? How are the processes run and controlled without this information? How are decisions made with regard to changing operational conditions without this information?

8   When chemists, chemical engineers, analytical chemists, and process managers discuss chemical reactions, do they all use the same "dictionary"? What are the consequences if terms such as conversion, yield, and selectivity are misinterpreted in defining process control and process measurements?

9   Are catalysts used in your process? How well is their exact role in the chemistry understood? If it were better understood, what be some advantages in terms of quality? Productivity? Are mechanisms for poisoning or reducing effectiveness of catalysts understood? How are decisions made regarding when and how to either regenerate or replace a catalyst? Science or history?

10   Are gases used as raw materials or produced as intermediates or products? If so, are their effects on physical properties within a reactor understood? Do these effects change with process conditions or within a process cycle? Is it known whether gases used or formed are ideal? If not, are the relationships between pressure, temperature, and the number of moles of gas understood? What are the consequences if this information is not known?

Review Questions (Answers in Appendix with Explanations)

1 Stoichiometry determines ratios and kinetics determine:
   A __Kinetic energy
   B __Rate
   C __Energy release
   D __Ratio of rate to energy

2 Competitive reactions refer to:
   A __Reactions that are also practiced by a competitor
   B __Multiple reactions that may occur from the same starting raw materials
   C __One or more reactions that compete for raw materials based on price
   D __One reaction that runs right after another

3 The same raw materials, combined in the same ratio, can produce differing products:
   A __Yes
   B __No
   C __Sometimes, depending upon value of the products produced
   D __Yes, depending upon reaction conditions

4 Thermodynamics of a chemical reaction determine:
   A __The amount of energy released or consumed (needed) if the reaction occurs
   B __Under what circumstances a reaction will occur
   C __Time delay in a reaction starting
   D __How dynamic the reaction is

5 A kinetic rate constant:
   A __Is affected by temperature
   B __Is not affected by stoichiometry
   C __Is not affected by altitude
   D __Is affected by size of reaction equipment

6 The rate of a chemical reaction:
   A __Can be changed by changing pressure and/or temperature
   B __Will be affected by stoichiometry and ratios of reactants
   C __Will be affected by how fast products are removed
   D __All of the above

7 The rate of a chemical reaction is typically _____ with temperature:
A __Linear
B __Quadratic
C __Logarithmic
D __Semilogarithmic

8 Conversion of a chemical reaction will always be:
A __The same or greater than yield of the same reaction
B __Less than the selectivity to multiple reaction products
C __Unaffected by the kinetic rate constant
D __Different from the selectivity of a reaction

9 If a calculated heat of a particular reaction is negative (exothermic), it means:
A __We don't want the reaction to occur
B __The heat calculation is incorrect as it should be a positive number
C __Energy is released if the reaction occurs
D __Energy is required to sustain the reaction

10 If a calculated heat of a particular reaction' is positive (endothermic), it means:
A __It is good for the reaction to occur
B __Constant energy input is required to sustain the reaction
C __The reaction will never stop once started
D __All of the above

11 Equilibrium in a chemical reaction system can be affected by:
A __Ratio of reactants
B __Temperature
C __Number of possible reactions
D __All of the above

12 The equilibrium constant K_e refers to:
A __The ratios of reactants to products
B __The ratio of reactants to products under certain conditions
C __The ratio of products to reactants
D __The ratio of products to reactants under specific conditions

13 A change in pressure will most likely affect reaction equilibrium for:
A __Liquid–liquid reactions
B __Liquid–solid reactions
C __Gas–gas, gas–liquid, or gas–solid reactions
D __Reactions using a gas whose price is increasing

14 The total time for a reaction to go to completion is affected by all of these except:

A __Kinetic rate constants

B __Rate of heat removal in an exothermic reaction

C __Stoichiometry of reactants

D __Size of the reactor

15 Catalysts can do these things:

A __Lower the temperature or severity of conditions of a reaction

B __Initiate an exothermic reaction

C __Favor one product over another in a reaction system

D __All of the above

16 The loss of catalyst effectiveness over time is most likely due to:

A __Change in stoichiometry in the feed

B __Poisoning or contamination

C __Change in catalyst vendors

D __The introduction of arsenic into the feed

Additional Resources

Fontes, E. (2015) "Modeling Chemical Reactors" *Chemical Engineering Progress*, 2, pp. 46–49.

Loffler, D. (2001) "Avoiding Pitfalls in Evaluating Catalyst Performance" *Chemical Engineering Progress*, 7, pp. 74–77.

Milne, D.; Glasser, D.; Hildebrandt, D. and Hausberger, B. (2006) "Graphically Assess a Reactor's Characteristics" *Chemical Engineering Progress*, 3, pp. 46–51.

Worstell, J. (2001) "Don't Act Like a Novice about Reaction Engineering" *Chemical Engineering Progress*, 3, pp. 68–72.

5

Flow Sheets, Diagrams, and Materials of Construction

The conceptualization of a full-scale process is usually done in stages. The exception to this might be the design and construction of a well-established process for producing a commodity chemical whose design has remained relatively unchanged over the years and may be in the hands of an engineering and construction contractor as an "off-the-shelf" plant.

Assuming this is a new process for scale-up, chemical engineers will work with a chemist to understand the product being made and the laboratory process by which it is made. The chemical engineer will watch the lab process being run to ensure an understanding of the reaction conditions and their limitations. Since it is unlikely that the full-scale process will look exactly like the lab process (and may become continuous vs. the batch process used in the lab), a simple qualitative flow sheet, frequently called a "block" flow diagram, will be developed showing the types of equipment that might be used and their relationship to one another. An example of such a flow sheet for a waste treatment process is shown in Figure 5.1.

This type of flow diagram is only qualitative and shows only the basic concept of how the process is envisioned to work. It shows the feeds (air and screened effluent), a clarifier/settler (without saying exactly how it operates or under what conditions), and the fact that some of the activated sludge is recycled in the process (but not how much). It also shows the end product and the treated effluent but says nothing about the composition or temperature. It also does not indicate how the air is actually introduced into the tank or under what conditions (pressure, temperature). This type of flow sheet is just a starting point for discussion and an initial thinking about how the chemistry will be practiced on an industrial scale. For example, the handling of raw materials will certainly be different. They will not come out of reagent grade bottles, but more likely from pipelines, rail cars, trucks, or drums. There will need to be storage facilities and safety systems in place to handle large quantities of raw materials. If there is a reaction vessel or system, the reactor may or may not be a stirred vessel, but the raw materials still need to

Chemical Engineering for Non-Chemical Engineers, First Edition. Jack Hipple.
© 2017 American Institute of Chemical Engineers, Inc. Published 2017 by John Wiley & Sons, Inc.

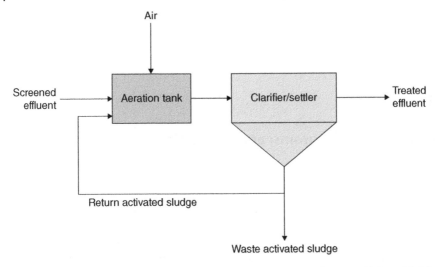

Figure 5.1 Simple block flow diagram. Source: Chemical Engineering Progress, 3/12, pp. 29–34. Reproduced with permission of American Institute of Chemical Engineers.

go in and the products and by-products still need to be separated and recovered (the carbon dioxide removal box in this diagram is an example of this). There may be separation processes needed, assuming that what comes out of a reactor is not the final product and needs to be separated and purified in some way.

As we obtain more knowledge of the chemistry and the process, we can then begin to add detailed information to the flow sheet such as flow rates, compositions, and temperatures. This will take the form of process flow diagrams and are known as a piping and instrumentation diagram, showing the details of the process and how each unit operation will be controlled, as shown in Figures 5.2 and 5.3.

Figure 5.2 shows additional details of the internal designs of the vessels, how the sludge will be recycled, preliminary indications of what will be measured and controlled, and an indication of the need for "backup" equipment (as in the pumps and blowers).

Figure 5.3 shows the detail of just one section of the process flow diagram and the manner in which it will be controlled.

As a subset to this level of diagramming and flow sheets, we now see 3D diagrams of much of the equipment, generated by CAD/CAM programs, as shown in Figure 5.4.

This level of detail, which was not possible without the 3D software tools now available, allows the chemical engineer, along with mechanical, piping, and instrumentation engineers, to view access points to equipment and

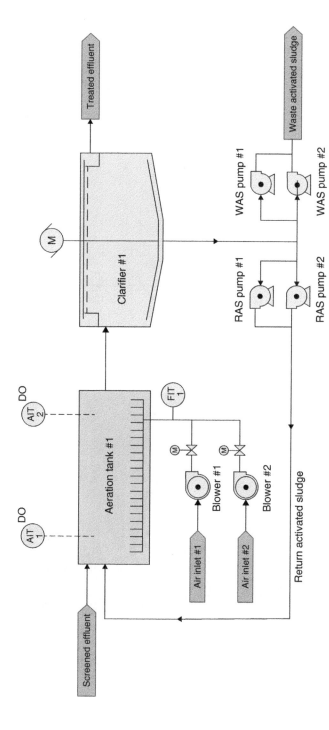

Figure 5.2 Process flow diagram. Source: Chemical Engineering Progress, 3/12, pp. 29–34. Reproduced with permission of American Institute of Chemical Engineers.

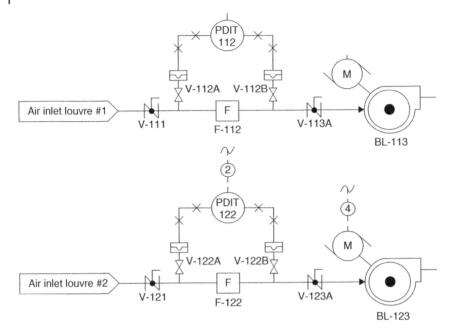

Figure 5.3 Detailed process flow diagram including instrumentation. Source: Chemical Engineering Progress, 3/12, pp. 29–34. Reproduced with permission of American Institute of Chemical Engineers.

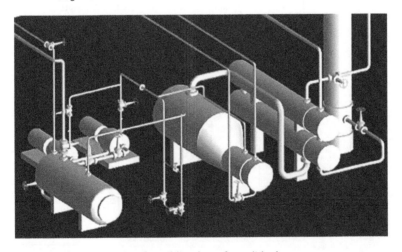

Figure 5.4 3D process view for public release from slide share.

instrumentation. This is important not only from a practical operational and maintenance standpoint but also from a safety standpoint. We need to understand how easy it is to access emergency valves and to access escape routes in emergency situations.

One of the challenges with any kind of flow sheet, whether it be on paper or in electronic format, is keeping it up to date. There must be a disciplined process for keeping these important records accurate and up to date in the face of minor process improvements, as well as maintenance and instrumentation changes. If an outdated flow sheet is used in a safety, HAZOP, or reactive chemicals review, the consequences can be serious.

Materials of Construction

When we build a large chemical facility, we will not be using the glass material that may have been used in the laboratory. Glass, though corroded by some materials such as fluorides and strong alkalis, is a very corrosion-resistant material. However, it is not pressure resistant, unless used as glass lined metal, and has very low shock resistance. If used on a large scale in a non-glass lined form, it presents the possibility of large-scale chemical releases.

Large plants will replace glass with the least costly metal or glass-lined metal material that will provide sufficient corrosion resistance. Experienced corrosion engineers are heavily involved in these decisions as there are many corrosion situations where the proper choice of materials goes against common sense. For example, chlorine (Cl_2) is known to be a corrosive compound, but if it is kept very dry, it can be handled in steel for an extended period of time. However, if the chlorine is wet or saturated with water, it will corrode normal steel extremely rapidly. Wet chlorine is typically handled in titanium piping. If dry chlorine is used in titanium, it will ignite and actually burn within the pipe ($2Cl_2 + Ti \rightarrow TiCl_4$). It is never a good idea to make the assumption that a more expensive metal is more corrosion resistant. Copper is also an unusual material in terms of its ability to resist corrosion to different classes of materials.

Another aspect of corrosion is the phenomenon of stress corrosion cracking. Materials such as stainless steel, having grain boundaries between their various phases, are subject to "short circuiting" along grain boundary paths, especially by chloride ions. This can appear as a catastrophic failure if a pipe "breaks" along its grain boundaries. This phenomenon can lead to catastrophic failure while a normal corrosion study may show very low corrosion rates. The grain boundaries in stress crack-prone materials can be seen in Figure 5.5, where the "short-circuit" paths are plainly visible.

A similar unusual behavior is "pitting" where holes develop in a material though its overall corrosion rate may be very low.

Corrosion testing and evaluation is typically done under closely controlled laboratory conditions, attempting to simulate actual process conditions. Basically, the sample of material of concern is inserted into the solution to which it will be exposed for a set amount of time, at a given temperature, and the weight of the sample compared with its original weight converted into loss of thickness of material and expressed typically in "mils/year," or

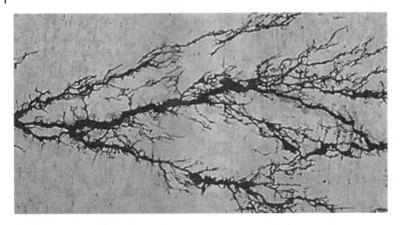

Figure 5.5 Stress corrosion versus general corrosion. Source: Reproduced with permission of NASA.

thousandths of an inch per year. A material with a corrosion rate of less than 5 mils/year would ordinarily be considered noncorrosive. Numbers greater than this will be classified as moderate or severe. The exact classifications are determined by industry standards set by organizations such as ASTM.

In many cases, it is not possible to provide a corrosion test that directly simulates a 30-year exposure to a given environment. We simply don't have that much time! In these cases, accelerated corrosion tests are frequently used, taking advantage of our knowledge of the basic chemistry of the corrosion chemistry, its rate constant, and the effects of temperature. For example, if we wanted to know the corrosion rate for water in steel at room temperature, we could raise the temperature of the water, and knowing how the corrosion reaction rate is affected by temperature, we take the data at a higher temperature and extrapolate it back to room temperature, obtaining the information in a shorter time.

Summary

Flow sheets are the visual way that chemical engineers show how a process is intended to run, the connections between the various unit operations, and the connections with raw material and utility supplies. In their advanced stages, they also show material and energy balances, recycle loops, and how a process is controlled. The visualization of the process, especially including a 3D perspective, can be an effective training and safety tool.

Coffee Brewing: Materials of Construction and Flow Sheets

Have you ever drawn a flow sheet or diagram of how you make a cup of coffee? Try it! What are the raw materials? What comes in? Goes out (spent grounds, waste pod, coffee not consumed)? In which order? Does it make a difference? What if you were asking someone else to make a cup of coffee, with a complicated recipe, for you? Of course it would! That's why we use flow sheets and the information on them to communicate how a process is run, connected, and controlled. What materials are used in the process? Most coffee carafes are glass, enabling us to see how much coffee is contained. We often see coffee residue (coffee "degradation" products) on these containers, despite having been through a washing machine. Carafes that insulate (as opposed to sitting on a hot plate) typically use a sealed vacuum layer. The metal is necessary to enable the construction of a vacuum layer. All three of these (clean glass, glass with residue, metal vacuum container) have different potential for product contamination, no different than the concerns we have in a chemical process for corrosion, cleaning, and impurities. More are discussed as we continue!

The materials used to construct a process must be chosen carefully and take into account not only normal operating conditions (temperature, pressure, quality/impurities in process streams) but also reasonable consideration for possible unexpected process conditions and impurities in raw materials. The materials that we use to construct the equipment and piping determine, to a great extent, how long the equipment and piping will last, the potential for corrosion products to cause product quality problems, and the original cost of the plant.

Discussion Questions

1 How up to date are your process flow sheets? What is the process for updating them? How would you know whether a flow sheet is up to date? Who would you ask?

2 When changes to a process are made, what is the mechanism for ensuring that these changes are transferred to the process and the plant records? How is it assured that the changes are accurate?

3 How are flow sheets used in safety and reactive chemicals reviews? How could they be used better?

4 How are the various levels of process flow sheets used in new employee training? How could they be used better?

5 Are the corrosion rates of the materials used in your process known? By whom? Where are the records kept? When tests are made, where are they recorded? How is the information communicated?

6 What actions are taken when a corrosion leak occurs in a process?

7 What possible contamination possibilities exist that might affect the corrosion resistance of currently used materials? How are these possible contaminants monitored?

8 What procedures are in place to prevent the use or installation of a "substitute/readily available" material or process component with less corrosion resistance?

Review Questions (Answers in Appendix with Explanations)

1 The level of detail contained in a flow sheet, in order of increasing complexity, is:
A __P&ID, mass and energy balance, 3D
B __Mass and energy balance, P&ID, 3D
C __Block flow, process flow, P&ID, 3D
D __3D, P&ID, mass and energy balance, block flow

2 Process flow diagrams are important because they:
A __Ensure disk space is used on a process control computer
B __Provide a sense of process stream and equipment interactions
C __Provide a training exercise for new engineers and operators
D __Make effective use of flow sheet software

3 3D process diagrams are most important because they:
A __Enable personnel to envision the interaction between people and equipment
B __Allow the use of 3D glasses from the movies that otherwise would be thrown away
C __Enable the use of 3D software
D __Show the best location for a security camera

4 It is important to ensure flow sheets are up to date because:

   A __They are used by maintenance personnel to identify connections and equipment

   B __They show safety valves and relief systems

   C __They are a means of common communication between engineers, operators, and maintenance personnel

   D __All of the above

5 Accurate measurement and knowledge of corrosion rates, as well as what affects them, within process equipment is important because:

   A __Pipe vendors need to know when to schedule the next sales call

   B __Corrosion meters need to be tested once in a while

   C __It is important to understand the estimated life of process equipment and the potential for corrosion products to contaminate process streams

   D __We need to keep evacuation plans up to date for equipment failures

6 A process fluid with higher water content than one with a lower water content:

   A __Will be more corrosive

   B __Will be less corrosive

   C __Depends on the temperature

   D __Can't tell without laboratory data

Additional Resources

Crook, P. (2007) "Selecting Nickel Alloys for Corrosive Applications" *Chemical Engineering Progress*, 5, pp. 45–54.

Gambale, D. (2010)"Choosing Specialty Metals for Corrosion-Sensitive Equipment" *Chemical Engineering Progress*, 7, pp. 62–66.

Geiger, G. and Esmacher, M. (2011) "Inhibiting and Monitoring Corrosion (I)" *Chemical Engineering Progress*, 4, pp. 36–41.

Geiger, G. and Esmacher, M. (2012) "Inhibiting and Monitoring Corrosion (II)" *Chemical Engineering Progress*, 3, pp. 29–34.

Hunt, M. (2014) "Develop a Strategy for Material Selection" *Chemical Engineering Progress*, 5, pp. 42–50.

Nasby, G. (2012) "Using Process Flowsheets as Communication Tools" *Chemical Engineering Progress*, 2, pp. 36–44.

Picciotti, M. and Picciotti, F. (2006) "Selecting Corrosion-Resistant Materials" *Chemical Engineering Progress*, 12, pp. 45–50.

Prugh, R. (2012) "Handling Corrosive Acids and Caustics Safely" *Chemical Engineering Progress*, 9, pp. 27–32.

Richardson, K. (2013) "Recognizing Corrosion" *Chemical Engineering Progress*, 10, p. 26.

Walker, V. (2009) "Designing a Process Flowsheet" *Chemical Engineering Progress*, 5, pp. 15–21.

6

Economics and Chemical Engineering

As we have discussed, there must be an incentive to commercially produce a product or service. This means that the price of the product must exceed all the costs of making the product, including a return on the capital invested to build the facility. In principle, this no different than the expectations someone has for a return from a bank deposit or stock market investment. This margin may vary to some degree with the cost of borrowing money over time, but it still must be greater than zero. For example, if we want to commercialize a reaction of the type $A + B \rightarrow C$, what should be considered and estimated?

What is the cost of A? B? At what volumes? At what purity? The value of C? What might affect these values? What is the cost of the arrow? Capital? Energy? Labor? Waste disposal? Ultimate equipment disposal? Ultimate site cleanup? Does the reaction go to 100% to the product desired? It is unlikely that "C" is the only product produced. What are the other materials produced? How are they separated from the desired product? Can they be recycled and reused? Separated and used as raw materials for other useful reactions? Are there waste disposal costs and issues? How hazardous are A, B, and C? What is the cost of ensuring that they are all handled and stored safely? What are the safety aspects of the arrow of conversion?

In addition to evaluating overall costs, there are two basic categories to be considered, fixed and variable. What is the difference? Fixed costs describe funds that are spent ahead of actually producing any product and, in addition, are independent, or nearly independent, of the rate of product produced (a plant may be designed to produce 1000 #/h, but actually may run and produce only 800 #/h so this would be 80% of capacity). Examples of such costs are the purchase and installation of the equipment in the plant, utility and waste handling facilities related solely to this process, purchase of land to site the facility (if it is not already owned), and insurance and taxes. In addition, there are costs in operating a plant that are nearly independent of its operating rate. Labor cost, in many situations, is relatively fixed as it is unlikely that a large personnel

Chemical Engineering for Non-Chemical Engineers, First Edition. Jack Hipple.
© 2017 American Institute of Chemical Engineers, Inc. Published 2017 by John Wiley & Sons, Inc.

reduction will occur in a continuous plant whose rate may vary. However, in a batch chemical manufacturing facility which reduces the number of shifts it operates, labor costs may be somewhat variable. Liability insurance and workman's compensation fees will also normally be considered a fixed cost. Basic supplies for plant facilities such as personnel facilities are also fixed. There are many situations where a state or county may provide tax incentives to a company to locate a facility and create jobs. This normally takes the form of eliminating or reducing local property tax costs for an extended period of time. It could also take the form of providing access streets and roads that would have an effect of reducing the total capital cost of building the facility. These incentives may also be impacted by the continuing number of jobs versus those created initially.

An important factor in evaluating fixed costs is the concept of depreciation. The US (and most other countries') tax code provides an indirect incentive for companies to build plants via this concept. Let's assume that a plant costs $100 M to be built. This is the capital cost referred to earlier. In general, the tax code allows a company to "depreciate" approximately 10% of this fixed cost every year and deduct this amount from its tax bill. This is a part of the federal tax code system and could be changed by the US Congress, either permanently or on a short-term basis for economic stimulation. We can look at this as a type of "forced" savings account that generates funds to rebuild the plant at a future date. The amount of this depreciation can be (and has been on occasion) changed by national and state legislatures to provide an incentive to build plants that provide jobs. This is shown as a "cost" on the company's accounting books when profits are being reviewed, but when a company is showing its "cash flow," the depreciation is added back in. Cash flow includes both profit and depreciation.

The cost of buying equipment is only one part of the fixed cost. The equipment must be installed. The cost of this will vary greatly with a number of factors including buildings needed to enclose the process, piping connections between equipment and process vessels, electrical supply to the equipment, foundation preparation, cost of supplying utilities (Are they already available? Are they coming from a city or public utility?), fireproofing required, cost of instrumentation, emergency backups, and contractor fees and margins. The last item will be greatly affected by conventional supply and demand economics and how busy the contractor is or wants to be!

Variable costs are those directly related to the rate at which the process operates. This would include raw materials and energy costs directly related to the process (e.g., as opposed to heating and lighting for the building in which the plant operates). Note that energy costs could be offset to some degree if a reaction generates heat that can be used elsewhere in a facility. Also included here would be environmental control or waste disposal/destruction costs. Typically these are proportional to the level of production.

In all cost estimates, a contingency is included. This can vary from a small percentage to account for minor unanticipated events during a plant start-up to a large percentage if the plant is the first of its kind using new chemistry that a company has no previous experience with.

The sum total of all these factors can vary from 3 to 7 times the cost of the purchased equipment. Unless this process is "first of its kind" on a bare plot of land, a number frequently used is 3.5–4 times the purchased equipment cost.

Distribution and shipping costs also come into consideration. Most chemical materials are sold on a cost plus freight basis, meaning that the customer pays the transportation costs (rail car rates, truck rates for drums, etc.). However, in many cases of large-volume commodity products such as chlorine, sulfuric acid, ammonia, and so on, we will frequently hear the term "freight equalized," meaning that the producer has partially compensated for the freight cost if there is another supplier physically closer to the customer.

Let's take a look at a hypothetical example using the following general chemical reaction:

$$A + B \rightarrow C$$

We will assume the following:

1) The reaction is endothermic (i.e., it requires constant supply of energy to proceed; without energy input, it stops), requiring 2000 BTU/mole of "C" produced.
2) The cost of "A" is $1/mole and "B" is $2/mole.
3) The value (selling price) of "C" is $5/mole.
4) The cost of the solvent, in which the reaction is run, is $0.50/mole, and we need 10 moles of solvent per mole of product. Most of the solvent can be recycled.
5) The cost of building a plant (installed cost) is $50 M (million).
6) The capacity of the plant is 5 M moles/year.
7) The depreciation rate is 10%.
8) The combined federal, state, and local taxes are 40% of net profit.
9) This is a new process and product for the company.
10) The process is a one-step process, using a catalyst under high pressure and a reaction solvent.
11) The product is a solid which precipitates from the reaction as it is formed. The reactants are soluble in the solvent; but the product is not. The product must be filtered and dried prior to storage and sale.
12) The conversion in the reaction is 100%, but the yield is only 90%, meaning that there is some unreacted raw material that must be recycled for reuse. This is done by recovering the solvent and unreacted raw material and recycling them.
13) One of the raw materials, as well as the reaction solvent, is flammable and hazardous.

An overall early-stage process flow diagram might look like that shown in Figure 6.1.

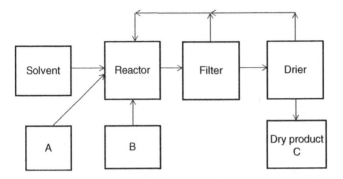

Figure 6.1 Simple block flow diagram for early-stage economic analysis.

The general points about this diagram are as follows:

1) Possible slow degradation of either the solvent or either of the reactants.
2) The storage and inventory requirements of any of the reactants, the solvent, or the product.
3) Materials of construction requirements.
4) The type of instrumentation required.
5) The effect of any physical property changes.
6) The nature of the pumps, the drier, and solids transport equipment.
7) This flow sheet also assumes that the unreacted raw material and the solvent can be recycled together.
8) The flow diagram shows no detail on the storage and handling of the final product.

Depending upon the volume of products made, the kinetics and batch reaction size, and the limitations of building and equipment, this type of process might actually be installed and run in two or more parallel reaction systems.

What would the variable costs look like? The raw material costs are $1 and $2/mole for A and B, respectively, and both are eventually converted to product, though "B" must be recycled. Since we need one mole of each, the total raw material cost is $3/mole of product. The other raw material that needs to be considered is the process solvent. It is not logical to assume that there will be no solvent loss, and environmental regulations may have a role to play in the decision as to the degree of recovery required.

The energy cost is at least the cost of energy addition for the endothermic reaction, stated at 2000 BTU/mole of product. But it is also stated that this is a high-pressure reaction, implying the need for moderate pressure steam. The thermal value of steam condensing is about 1000 BTU/pound, so if we assume that this energy is supplied by steam to the jacket of the reaction vessel and that the steam costs are $4/M BTU at a higher than normal pressure, this amounts to $4/1 000 000 = 0.00004/BTU × 2000 BTU/mole of product or $0.008/mole of product. To this we might add utility costs related to hot and cold water for restrooms, safety showers, and process equipment wash-downs, estimating the total at around $0.01/mole of product. There will also be some very minor cost associated with office space, external maintenance, and other miscellaneous items, but, compared with the three costs we have considered, are most likely insignificant. So here is the summary of the approximate variable costs:

Variable costs for product "C"

Raw materials ($1 + $2)/mole	$3.00/mole of product
Solvent (1% loss) 10 moles/mole of product at 0.50/mole × .01	0.05/mole of product
Energy	0.01/mole of product
Total variable costs	$3.06/mole of product

Let's now take a look at the fixed cost side. The cost of building this plant is $50 M to produce 5 M moles/year, so the capital cost, on a ratioed basis, is $10.00/mole. Assuming that we are allowed to depreciate this in the normal 10%/year fashion, this amounts to $10.00/10 or $1.00/mole.

Maintenance is figured as a percentage of capital costs, and we are told this is a high-pressure reaction system, so it is likely to be on the higher side, possibly 8% of capital, or 0.08 × $50 M/5 or $.80/mole.

Labor, unless the plant is temporarily shut down, will normally use the same number of employees, even if the product rate is reduced slightly. Let's assume there are two operators per shift in a 24 h operation. This is the equivalent of eight full-time employees. Assuming that this is a normal Fortune 500 company with the usual benefits, this would probably cost the company about $80 000/employee or $640 000/year or $640 000/5 000 000 or $0.13/mole.

There will be other fixed costs such as local taxes, liability, and workmen's compensation insurance which tend to be rather minor (per pound or mole). On the fixed capital side, we are assuming the availability of utilities such as moderate pressure steam, water, and electricity for lights and pumps. If this were a "grassroots" facility, additional capital would be required.

The following is the summary of the capital costs:

Fixed costs for product "C"	
Depreciation	$1.00/mole of product
Solvent (1% loss) 10 moles/mole of product at 0.50/mole × 0.01	0.05/mole of product
Energy	0.01/mole of product
Total variable costs	$1.06/mole of product
Adding back in the variable costs, we see the total picture:	
Variable costs	
Raw materials ($1 + $2)/mole	$3.00/mole of product
Solvent (1% loss) 10 moles/mole of product at 0.50/mole × 0.01	0.05/mole of product
Energy	0.01/mole of product
Total variable costs	$3.06/mole of product
Total costs	$4.12/mole of product

At a selling price of $5.00/mole, that leaves a gross profit of $5.00 minus $4.12 or $0.88/mole.

What is the return on investment (ROI) before taxes? It's the profit divided by the capital invested – no different than how we make decisions about where to invest our money. Our profit is $0.88/mole × 5 000 000 moles/year or $4 400 000/year. Dividing this by the investment made gives the ROI before taxes of $4 400 000/$50 000 000 or 8.8%. For a moment, think about this in the context of your own personal investment decisions, before taxes. Would you make this investment? For most chemical companies, unless this is a carbon copy of a plant built and run many times before, 8.8% would be below the minimal acceptable return, given the risk of running a chemical plant versus investing in a long-term bond, for example. Now we have to pay taxes at 40%, equal to 0.4 × $0.88 or $0.352/mole. At a production rate of 5 M moles/year, this amounts to 5 000 000 × $0.352 or $1 760 000/year.

The after-tax profit is the after-tax income divided by the investment or $1 760 000/50 000 000 or 3.5%. Is this worth doing? In general, the answer would be no. Again, it's possible to get this level of return on a long-term bond of some sort.

So what would chemical engineers consider if they were reviewing this project and its economics and were charged with making some positive input to the R&D program associated with it?

1) Seventy-five percent of the total cost is in raw materials. What is the basis for the numbers used? Were they just taken from a trade magazine or from a substantive conversation with a purchasing agent? What is the purity of the raw materials assumed? If the prices in this review were at 99.9% purity, what is the change in price if 99% were acceptable? What would have

to be changed about the process to accommodate this lower-quality raw material? How would this impact the capital costs?

2) The reaction, as currently described, does not use a catalyst. Has this been considered? If the reaction rate could be enhanced (e.g., lower temperature, lower pressure), how much less capital would be required? If this were possible, how does the raw material cost change?

3) It is stated that this is an endothermic reaction requiring heat input. Is there another process nearby, using exothermic chemistry, which could be coupled in some way to provide the energy?

4) What could be changed about the composition, purity, or form of the product which would allow a higher price?

These are just a few of many questions that should be asked at an early stage of such a project.

Summary

The four primary issues in determining the profitability of a chemical process are costs of raw materials, cost of building and maintaining the plant, cost of utilities (energy, water, and electricity), taxes and depreciation, and selling price of the product being manufactured. The impact of changes and variability in the variable cost side of the total is critical. These numbers and what underlies them can also serve as a foundation for long-term process R&D activities to reduce the most significant costs and to look at less expensive raw material sources tied to new chemistry. Fixed costs, once incurred, are spent and cannot be retrieved without a "write-off" of the facility.

Coffee Brewing: Cost

Does all coffee cost the same? Of course not! But why? What is the cost of the raw material (the beans)? How rare are they? From what country? How easy is it to harvest, inventory, and ship? From where? The various processes to produce and package beans, evaporation to produce instant coffee, vacuum to produce freeze-dried coffee, grinding coffee, ground coffee in a vacuum container (vacuum is not free), and shipping/distribution/storage all have different costs associated with them. This must be weighed against what we think the consumers are willing to pay for all these different options. There are indirect costs associated with specialized coffee brewing at home. It is in the form of time which indirectly has a cost associated with it. After a pot of coffee is made, a significant portion of it may be thrown away. How much does this cost? The new "K-Cups" avoid this, but the cost of the coffee in them is significantly higher than that in cans or bags. What is this convenience worth versus the cost of making these very small packages? More later! No rights or wrongs...just choices based on criteria set ahead of time.

Discussion Questions

1   Is the split between fixed and variable costs for all of your processes well understood? How is this ratio affected by % capacity the plant is running?

2   Does the process R&D prioritization reflect that difference?

3   What operating plans are in place for a significant decrease or increase in product demand?

4   What possible major shifts in raw material costs and supply might be anticipated? What resources are dedicated to scouting for possible paradigm shifts in this area?

5   What possible changes to tax and depreciation laws may be on the horizon? In what areas? Who has the responsibility to monitor these and make recommendations?

6   What variables determine your organization's "acceptable" return on investment (ROI) criteria? How were they determined? Are these criteria reviewed? How often? By whom? What are the criteria?

7   To the extent that utilities such as power and water are critical parts of the economics and reliability of the process, what kinds of communication are in place with the utility supplier? What are their long-term plans? Do they know about yours, or do you just assume that whatever you need will be there when you need it? If a utility merger or acquisition involving your utility supplier is considered a possibility, how might it affect you?

8   How would a severe drought affect you? Your utility supplier? What backup and alternatives have you considered and planned for?

Review Questions (Answers in Appendix with Explanations)

1   The cost of manufacturing a chemical includes:
    A __Capital cost (cost of building the plant)
    B __Cost of raw materials
    C __Taxes, labor, supplies
    D __All of the above

2 The most important factors in determining the variable cost of manufacture are typically:
   A __Shipping costs
   B __Labor contract changes
   C __Raw material and energy costs
   D __Security

3 If capital costs are 50% of the total cost of manufacture, and the production rate is reduced by 50%, the impact on the product's cost of manufacture will be:
   A __10%
   B __25%
   C __50%
   D __75%

4 If the cost of one raw material, representing 20% of a product's total cost, is raised by 25%, the impact on total cost will be:
   A __24%
   B __4.4%
   C __5.4%
   D __6.4%

Additional Resources

Bohlmann, G. (2005) "Biorefinery Process Economics" *Chemical Engineering Progress*, 10, pp. 37–44.

Burns, D.; McLinn, J. and Porter, M. (2016) "Navigating Oil Price Volatility" *Chemical Engineering Progress*, 1, pp. 26–31.

Moore, W. (2011) "'Lowest-Cost' Can Cost You" *Chemical Engineering Progress*, 1, p. 6.

Nolen, S. (2016) "Leveraging Energy Management for Water Conservation" *Chemical Engineering Progress*, 4, pp. 41–47.

Swift, T. (2012) "New Chemical Activity Barometer Signals Future Economic Trends" *Chemical Engineering Progress*, 8, p. 15.

Swift, T. (2012) "Energy Savings through Chemistry" *Chemical Engineering Progress*, 3, p. 17.

7

Fluid Flow, Pumps, and Liquid Handling and Gas Handling

This will be the first of the chapters related to the technical aspects of chemical engineering. As stated earlier, we tend to group chemical engineering studies and analyses by the term "unit operations." Fluid flow will be the first of these to be discussed.

Let's first define what we mean by a fluid flow. It is the flow of material through a *bounded* region (i.e., pipe) as opposed to the flow of a liquid or gas in an uncontrolled way, such as a vapor release or leaking tank. The behavior and properties of fluids affect how pumps, reactors, compressors, valves, and relief valves are designed and operated.

Fluid Properties

There are a number of fluid properties that affect how fluids behave as well as how much energy is required to move them:

Density is the weight per unit volume of a fluid. Water has a density of $62.4 \#/ft^3$ or $1000 \, kg/m^3$. There is a widespread in various fluid densities, as seen in Table 7.1 with approximate density values.

Table 7.1 Densities of common fluids.

Gasoline	$720 \, kg/m^3$
Liquid ammonia	$682 \, kg/m^3$
Liquid chlorine	$1442 \, kg/m^3$
Liquid bromine	$3100 \, kg/m^3$
Crude oils	$816–880 \, kg/m^3$
Milk	$1035 \, kg/m^3$
Blood	$1065 \, kg/m^3$
Mercury	$13500 \, kg/m^3$

Source: Public literature sources, averaged.

Chemical Engineering for Non-Chemical Engineers, First Edition. Jack Hipple.
© 2017 American Institute of Chemical Engineers, Inc. Published 2017 by John Wiley & Sons, Inc.

The density of gases is several orders of magnitudes less. The density of air at atmospheric conditions is around $1.25\,kg/m^3$. Increased pressure will raise these numbers. Some approximate additional gas densities are shown in Table 7.2.

Table 7.2 Density of common gases.

Ammonia	$0.73\,kg/m^3$
Chlorine	$3.20\,kg/m^3$
Bromine gas	$4.4\,kg/m^3$
Propane	$1.9\,kg/m^3$
Methane	$0.7\,kg/m^3$
Carbon dioxide	$1.9\,kg/m^3$
Steam	$0.59\,kg/m^3$
Hydrogen	$0.09\,kg/m^3$

Source: Public literature sources, averaged.

.   The higher the density of a fluid, the more energy is required to move or pump it. The densities of most fluids and gases commercially used are readily available online, in handbooks, or from manufacturers of the materials.

Viscosity is an important physical property, often overlooked in fluid flow calculations. It is stated in a number of ways. The *dynamic viscosity* is a measure of a fluid's resistance to flow. We can envision this with a fluid between two solid surfaces and we move one of the surfaces. How much force does it take to cause the fluid to move? It is basically a measure of a fluid's resistance to shear and has the units of #/ft s. 1 #/ft s. is one *poise*. It is represented in equations and calculations by the symbol μ. Many common fluids have viscosities in the range of 1 centipoise (0.01 poise), normally designated as 1 cP. As opposed to density, the viscosity of a fluid changes dramatically (orders of magnitude in some cases) with temperature. As a reference, the viscosity of water at room temperature is about 1 cP. Table 7.3 gives some ranges of viscosities of some common fluids.

Table 7.3 Dynamic viscosities of common fluids.

Gasoline	0.4–0.9 cP
Liquid ammonia	0.27 cP
Liquid chlorine	0.33 cP
Conc. Sulfuric acid	24 cP
Motor oils	65–315 cP
Molasses	5–10000 cP
Blood	3–4 cP
Catsup	50–100000 cP

Source: Public literature sources, averaged.

An increase in viscosity will make the liquid more resistant to flow (and thus harder to pump) and the converse is true. As opposed to density, viscosity is greatly affected by temperature. Table 7.4 gives the data for water.

Table 7.4 Viscosity of water versus temperature.

Water viscosity (cP)	Temperature (°C)
1.31	10
1.00	20
0.65	40
0.55	50
0.40	70
0.32	90
0.28	100

Source: Public literature sources, averaged.

In general, the viscosity of a fluid is logarithmic with temperature, as shown in Figure 7.1.

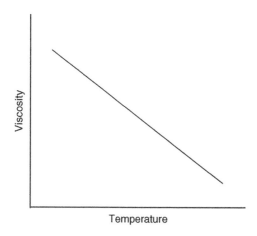

Figure 7.1 General response of viscosity to temperature.

This graph mirrors the dramatic drop in viscosity seen in the tabular data.

Another factor that affects actual viscosity is the response of viscosity to shear. In other words, how does a fluid's viscosity change (if at all) to the effect of the fluid being put under shear? Examples would include agitation, pumping, or mixing. There are four different classes of fluids:

1) Newtonian. A Newtonian fluid is one whose viscosity changes little with shear. Water, gasoline, and alcohols are examples of such fluids. Remember it still responds to a change in temperature!

2) Thixotropic. A thixotropic (shear thickening) fluid is one whose viscosity *decreases* with increasing shear but then returns to its higher viscosity state when the shear is removed. Examples include many inks (flow when applying but then stay where they're put), as well as paints, solder pastes, and quicksand.
3) Dilatant. A dilatant fluid is one whose viscosity increases with shear. Examples include a corn starch/water suspension or a transmission fluid.
4) Bingham plastic. This is an unusual fluid type whose viscosity versus shear curve is similar to a Newtonian fluid but not until a certain stress level is reached. In other words, the fluid is semisolid until a certain level of shear is applied, and then it begins to flow as a Newtonian fluid. Examples include drilling fluids and muds and catsup (ever tried to get catsup to flow out of a bottle?). A graphical display of these different classes of fluids is shown in Figure 7.2.

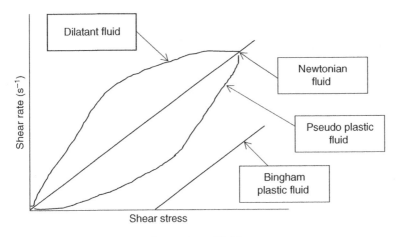

Figure 7.2 Shear versus stress for a variety of fluid types.

From this overview, it is obvious how important it is, when designed a fluid handling system, to be aware of the types of fluids being handled as well as well as their response to process conditions.

Kinematic viscosity is the ratio of dynamic viscosity to density, thus taking into account the density of the fluid.

The addition of solids to a liquid (creating a slurry) adds additional complexity to fluid behavior and analysis. The addition of solids can change the shape of the response curve for shear versus stress and needs to be carefully measured over the entire range of solids expected to be handled. In most cases, the addition of solids, creating a slurry, will increase the solution's viscosity. The exact change will be affected by solids concentration, particle size, and nature of starting liquid's viscosity.

As opposed to liquids, gas viscosities *rise* with temperature and generally follow an equation of the following type:

$$\mu = \mu_0 + \alpha T + \beta T^2$$

These constants, alpha and beta, must be experimentally determined or found in literature sources. Gas viscosity is also affected by pressure, whereas in liquid systems, this is not usually the case. Since processing of gases is very common in the hydrocarbon area and there is a large range of pressures and temperatures, it is important to have the knowledge of the gas viscosity response to temperature. Table 7.5 shows this type of data for air.

Table 7.5 Air and water viscosities versus temperature.

Temperature (°F)		0	20	60	100
Air viscosity	# s/ft^2	3.38	3.50	3.60	3.94

Used with permission of engineeringtoolbox.com

Characterizing Fluid Flow

One of the key ways we characterize a fluid is how *turbulent* its flow is in a pipeline (there are some analogous ways of analyzing turbulence in a vessel, which we will cover in Chapter 16). If a fluid is not mixed thoroughly within a pipeline, there is insufficient kinetic energy to overcome the fluid adhesion to the wall of the pipe; if the velocity is high enough, the flow (kinetic) energy is sufficient to overcome the fluid/wall adhesion. If we were to look at the velocity profiles across the pipe for these two types of flow, we would see profiles as shown in Figure 7.3. The linear profiles shown for turbulent flow are never an exact straight line as there is always a small amount of wall adhesion no matter what the velocity of the fluid is.

The flow with uneven distribution of flow velocity is called *laminar* flow, while the well-mixed flow distribution at a higher velocity is called *turbulent* flow.

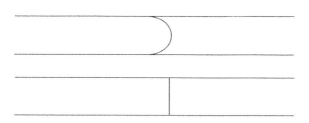

Figure 7.3 Velocity profiles: laminar versus turbulent flow.

Velocity is not the only parameter affecting which flow domain we are in but also the liquid physical properties and the diameter of the pipe. We can calculate what we call the *Reynolds number* for any fluid in a pipe with this equation:

$$N_{\text{Re}} = \frac{DV\rho}{\mu}$$

Where D is the pipe diameter, V is the velocity in the pipe, ρ is the density of the fluid, and μ is the liquid viscosity. Remember that we need to be consistent in the units we use in this calculation. This is a *dimensionless* number (and there are many in chemical engineering). In other words, if we plug numbers and units into this equation, we should have a number without any units. For example, if we have a fluid of 60 #/ft^3 density and a viscosity of 0.002 #/ft s. (2 cP) flowing at a rate of 5 ft/s in a ½ ft (6" pipe), we would calculate the Reynolds number to be

$$\frac{0.5 \text{ ft.} \times 5 \text{ ft./s} \times 60 \text{ #/ft.}^3}{0.002 \text{ #/ft. s}}$$

or 75 000 (and that number would have no units as all unit values would cancel out in the equation). We know from our experiments with flow in glass pipe studies that a Reynolds number below 2 000 implies laminar flow while a Reynolds number above 4 000 ensures turbulent flow. In between these numbers, things are not so clear and additional study may be needed to determine the flow regime. If we just look at the Reynolds number equation, we can think through some practical aspects of flow:

1) Increasing velocity will increase turbulence and the Reynolds number (but also increase pressure drop). Decreasing pipe diameter will have the same effect.
2) Increasing the density of a fluid will decrease the velocity and Reynolds number. Higher density fluid takes more energy to move.
3) Increasing viscosity (think maple syrup vs. water) will decrease the Reynolds number and velocity, thus making it more difficult to mix in a pipeline.

An additional general area of importance in sizing piping and pumps is the pressure drop through process piping. When process piping is new, there are no special pressure drop concerns other than to do the initial calculations correctly. However, over time additional friction loss can develop due to corrosion or buildup of materials on the inner surface of process piping.

This additional friction loss is accounted for by what is known as the Fanning friction factor, or "f." It is basically the percentage increase in pressure drop over time, determined through knowledge of the Reynolds number (degree of turbulence) and the relative roughness of the inner pipe surface. This can change over time due to corrosion products and contamination. These two factors are plotted in logarithmic fashion against each other, and we see a

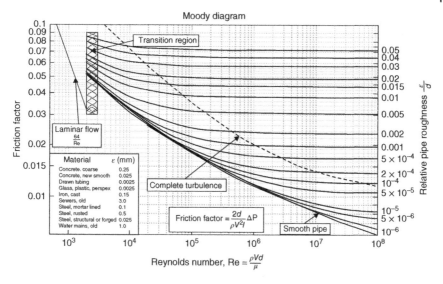

Figure 7.4 Friction factor versus Reynolds number. Source: Beck and Collins, https://commons.wikimedia.org/wiki/File:Moody_diagram.jpg. Used under CC BY-SA 3.0, https://creativecommons.org/licenses/by-sa/3.0/deed.en

logarithmic estimate of the friction factor that needs to be added to take into account the anticipated loss of pipe diameter. As seen in Figure 7.4, the net results are typically in the area of a few percent and are shown on a Moody diagram. Whether this is important or not depends upon how close the design was initially, the type of piping material, and the surface roughness. Figure 7.4 also shows some relative roughness factors for such materials as concrete, glass, and steel.

Figure 7.5 shows a simple pumping system moving a fluid from one tank to another.

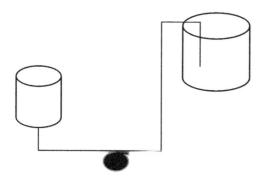

Figure 7.5 Pumping from one tank to higher level tank.

Review Figure 7.5 and make a list of all the pressure drop causes that would have to be taken into account in sizing the pump.

Sources of pressure drop:

1. _____

2. _____

3. _____

4. _____

5. _____

6. _____

7. _____

8. _____

9. _____

10. _____

We should consider, first of all, the level differences between the two vessels. Second, the rate at which we desire to pump the liquid (e.g., gal/min or GPM). Third, the total length of the process piping. Fourth, each time the piping makes a directional change, there is an added pressure drop. In this case there are four elbows that will add to the pressure drop. This diagram has no valves or in-line measuring devices, but if it did, they would also be major sources of pressure drop.

We would also consider the fluid properties. This would include its density and viscosity, both of which will affect the pressure drop significantly. If these properties changed over time, the pressure drop would also change with time.

Pump Types

There are two general classifications of fluid pumps: centrifugal and positive displacement. Within these two general classifications, there are many subsets of design.

If we were to open up a centrifugal pump and view it in a cross section, we would see something similar to that shown in Figure 7.6.

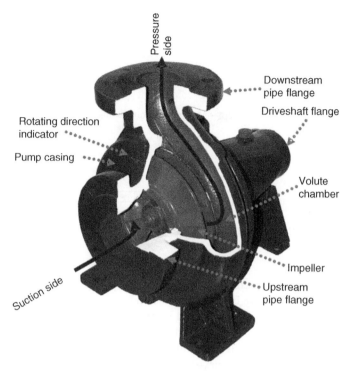

Figure 7.6 Internal view of a centrifugal pump. Source: Fantagu, https://commons. wikimedia.org/wiki/File:Centrifugal_Pump.png

The rotation of the blades converts the electrical power supplied to the pump to a "velocity head" or its equivalent pressure. The impeller is the rotational device internal to the pump, which rotates and transfers the motor's electrical energy into rotational energy. This type of pump does NOT produce constant flow under all conditions; it supplies a certain amount of energy via its rotational energy. The "head" for all fluids will not be the same. For example, the head or pressure produced by this type of pump will be higher for a low density fluid and lower for a high density fluid. Head and pressure are interchangeable and are related by the density of the fluid.

Any particular centrifugal pump will have what is referred to as a *characteristic curve*, supplied by its manufacturer, and it will look similar to that shown in Figure 7.7.

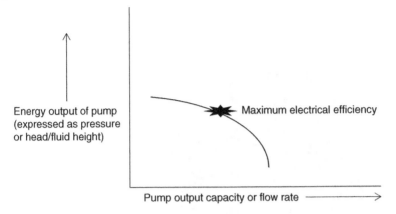

Figure 7.7 General pump curve.

As we can see, the head (or pressure) produced by this pump is inverse to its output. It is not possible to have both high flow and high head or pressure output with this type of pump. The pump will operate anywhere along this curve by controlling or varying process conditions, and the vendor usually indicates on the curve the highest electrical efficiency point for the pump's operation. At the extremes of this curve, maximum flow is achieved at minimum pressure or head output, and maximum flow is seen at minimum head or pressure output. It is not necessary to operate the pump here, but it is best to come as close to it as possible from an energy efficiency standpoint. This curve is unique to this pump with a specific size impeller (this is the diameter of the circulating veins in the pump shown earlier) in it. The shape of this curve can be shifted by changing the impeller size, internal to the pump, as shown in Figure 7.8.

The higher the flow or head generated by the pump, the more horsepower is required to drive it. Let's take a look at the centrifugal pump from an energy use standpoint (see Figure 7.9).

The hydraulic power needed by any given pump, P_h, will be a function of the flow rate (q), density of the fluid (ρ), and the differential height (including other pressure drop). "g" is the gravitational conversion constant in the units required. Note again the previous comments about unit consistency to ensure accurate calculations:

$$P_h(kW) = \frac{q\rho g h}{3.6 \times 10^6}$$

where P_z, the electrical power supply, is the $P_h/0.746$.

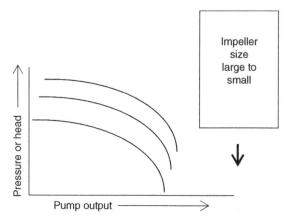

Figure 7.8 Pump output versus impeller size.

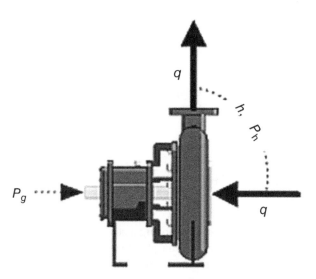

Figure 7.9 Energy use in a centrifugal pump. Source: Reproduced with permission of Engineering Toolbox.

Figure 7.10 shows a general representation of power requirements across a broad range of conditions.

The point of this analysis is not to memorize the equations (unless you are in the pump business) but to understand the variables that can affect a pump's performance. We can see, for example, that if a fluid's density is increased, the power requirements will increase proportionally, as will an increase in height difference.

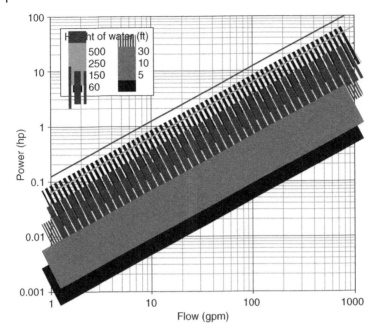

Figure 7.10 Shaft HP versus flow and height. Source: Reproduced with permission of Engineering Toolbox.

Net Positive Suction Head (NPSH) for Centrifugal Pumps

Another key mechanical design parameter in choosing a centrifugal pump for a particular application is to ensure that the pump's NPSH is at least what the system requires. NPSH is the minimum suction head (the sum of all the hydraulic pressure drops *ahead* of the pump's suction minus the vapor pressure of the fluid being pumped). Let's take the example of a tank holding 10 ft of liquid having a density of 0.8 (compared to water) and the equivalent of 2 ft of piping pressure drop between the tank and the pump. Let's also assume that the fluid in the tank is at a temperature where its vapor pressure is the equivalent of 1 ft of liquid ($1/14.7 \times 0.8$) feet. This means that the head available to the pump is $(10 \times 0.8) - 2 - 0.05 \times 0.8$ or just under 6 ft. The pump purchased for this service must be able to function on this amount of suction head. Otherwise, the energy supplied to the pump will be sufficient to boil the fluid in the pump, causing severe mechanical damage to the pump, including severe vibration and the pump potentially coming off its support base. Every pump purchased from a vendor will include this information. Because of unique and sometimes proprietary pump designs, it is possible for different pumps in the same general service area to require different NPSHs.

There are two frequently overlooked aspects of centrifugal pump design that occur. The first is a simple change in tank levels, either on the intake or discharge side of the pump, changing the pump output and its efficiency or allowing the operating conditions to go below the minimum NPSH required. The second is ignoring changes in the temperature of the fluid. An increase in fluid temperature feeding the pump will reduce the inlet pressure and its output and energy efficiency and affect the NPSH available. Fluids inside pumps with insufficient NPSH will boil (energy balance: the energy going into the pump needs to go somewhere), potentially causing significant stress and possible breakage of piping connections.

Positive Displacement Pumps

These are pump designs that generate a fixed amount of *flow* relatively independent of suction head available. These are usually piston or diaphragm pumps, which act on the principle of positive volume displacement versus generation of pressure or head. A gear pump is an example of such a pump:

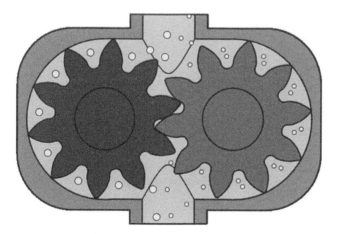

Figure 7.11 Gear pump. Source: https://commons.wikimedia.org/wiki/File:Gear_pump_animation.gif. Used under CC0 1.0, https://creativecommons.org/publicdomain/zero/1.0/deed.en. © Wikipedia.

Another example would be a diaphragm pump (Figure 7.12). The two diaphragms pulsate out of phase 180°, so if more continuous flow is desired, several of these pumps must be installed and operated out of phase with each

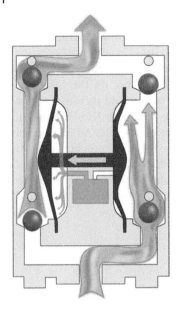

Figure 7.12 Diaphragm pump. Source: Samtar, https://commons.wikimedia.org/wiki/ File:Diaphragm_pump_animated.gif. Used under CC BY-SA 4.0, https://creativecommons.org/licenses/by-sa/4.0/deed.en. © Wikipedia.

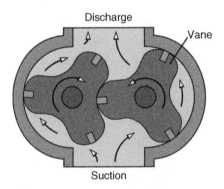

Figure 7.13 Rotary lobe pump. Source: https://commons.wikimedia.org/wiki/ File:LobePump_en.svg. Used under CC0 1.0, https://creativecommons.org/publicdomain/ zero/1.0/deed.en. © Wikipedia.

other. In order to have this type of pumping system deliver a more constant flow, one of more of them must be out of phase with each other to fill in the gaps and even out the flow.

A rotary lobe pump is another example of such a pump (Figure 7.13). There is one special safety aspect of positive displacement pumps that needs to be considered, and that is that they have the capability, since they are delivering constant flow, of creating vacuum on the suction side of the pump and the feed vessel must be designed to survive under vacuum or some kind of safety system must be in use to prevent a vacuum from being created. Also, since there is constant delivery of pressure output, the piping and all downstream equipment must be capable of handling the maximum potential pressure output of this kind of pump.

Variable Speed Drive Pumps

Another approach to varying and controlling the flow output from the pump is to use a variable speed drive motor, instead of constant speed motor whose output is typically throttled by a control valve on either sides of the pump. Their typically higher capital cost may be offset by energy savings in situations where there is constant valve cycling to control flow.

Water "Hammer"

Regardless of the type of pump being used, a general safety rule is not to make sudden changes in valving downstream of the pump, especially with positive displacement pumps. If we shut a valve suddenly downstream of the pump, preventing flow, the energy being put into the fluid must go somewhere (energy balance) and in this case will cause large vibration in the piping system possibly causing pipe breakage and, in rare cases, cause the pump to leave its support stand—all very dangerous situations.

Piping and Valves

The pump is only one part of a fluid handling system. Other parts include piping, valves, metering, and instrumentation. The diameter and type of piping (metal, plastic, plastic lined, etc.) will be chosen based on cost, corrosion properties, and pressure drop, and as mentioned several times already, there will seldom be one right choice. Piping is rarely simply just a straight line of pipe of one size in one direction. There will be elbows, joints, and connections of various types. Examples include tees (one line splitting into two), elbows (change in direction), couplings, reducing couplings, unions (straight line connection to allow access), and caps and plugs to allow access. From a practical standpoint, these are chosen for maintenance reasons, but each has a different impact on flow pressure drop and must be taken into account in sizing the pumping system.

Valves are chosen to control or modify flow and can be gate valves, butterfly valves, globe valves, and others. We will discuss these in more detail in our chapter on process control. But as with piping and valves, the pressure drop across valves must also be considered in the design of the liquid handling system and the choice of the pump.

Flow Measurement

There are many different ways of measuring flow, either as a mass unit or as a volume unit. If the density of the fluid is known to be constant, then volumetric flow measurement may be sufficient. In this case, there are several types of measuring devices commonly used.

One of the most commonly used is the orifice plate as shown in Figure 7.14.

The flow in the pipe goes through a fixed plate with a hole in it within the piping system. Based on the density of the fluid, the ΔP across the hole, and the ratio of the plate hole size to the pipe size, the volumetric flow rate can be calculated. The pressure in the pipeline will eventually recover. Another approach is to use a venturi meter as shown in Figure 7.15.

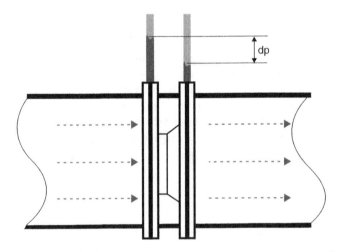

Figure 7.14 Orifice flow meter. Source: Reproduced with permission of Engineering Toolbox.

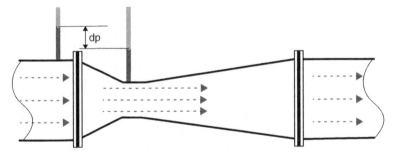

Figure 7.15 Venturi meter. Source: Reproduced with permission of Engineering Toolbox.

This type of meter operates under the same principle as the orifice meter, but since its geometry reduction is gradual on both sides of the ultimate hole, its pressure drop is less and the measurement is more accurate. But as you might expect, the cost is higher. Again, choices!

When it is important to know the actual mass (as opposed to volume) of flow, a Coriolis meter is frequently used:

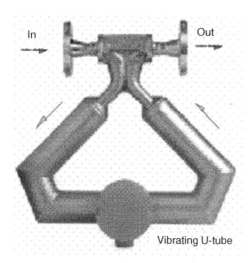

Figure 7.16 Coriolis meter. Source: https://commons.wikimedia.org/wiki/File:Coriolis_meter.png. Used under CC0 1.0, https://creativecommons.org/publicdomain/zero/1.0/deed.en. © Wikipedia.

This is a meter with a fixed amount of fluid volume going through it, but two parallel tubes containing liquid also vibrated at a fixed frequency. The change in frequency caused by a change in density allows this meter to compensate for this change and report out actual mass flow.

Gas Laws

When our systems involve the handling of gases, there are some unique differences in thinking about physical properties. Gases, unlike solids and liquids, change volume as a function of pressure. Solids and liquids do also, but to such a small degree, that it is usually not relevant in a practical sense. The equation used to describe the behavior of *ideal* gases is

$$PV = nRT$$

where
P is the absolute pressure of the system
V is the volume occupied by the gas
n is the number of moles of gas present
R is the universal gas constant
T is the temperature (in absolute, K or °R)

As we have discussed earlier, it is critical to use the value of R to match the units of the other variables.

Notice that the amount of gas present is represented by "n," the number of moles, not the mass or weight of the gas. That's because, as counterintuitive as it may seem, the volume is independent of mass and depends only on the number of moles. A gram mole of hydrogen, 2 grams, occupies the same volume as one gram mole of chlorine, 71 grams.

This simple law allows us to calculate and conceptualize the effect of changes in pressure and temperature. This equation applies only to an ideal gas, or as an approximation to a real gas that behaves sufficiently like an ideal gas. There are in fact many different forms of this gas law reflecting both molecular size and intermolecular attractions, and it is most accurate for monatomic gases at high temperatures and low pressures. The neglect of molecular size becomes less important for lower densities, that is, for larger volumes at lower pressures, because the average distance between adjacent molecules becomes much larger than the molecular size. The primary point here is that gases are different than liquids and gases, and chemical engineering calculations involving gases must take into account changes in pressure and volume.

Recalling Le Chatlier's principle, we can see how, in a system involving gases reacting or being generated, the equilibrium of the reaction changes by changing pressure. For example, if we have four moles of gases reacting generating two moles of gases, such as in the ammonia reaction

$$N_2 + 3H_2 \rightarrow 2NH_3$$

we would assume that a higher pressure during the reaction would shift the reaction to the right, increasing the amount of ammonia formed. This is in fact the case, as the ammonia reaction is usually run at pressures greater than 1000 psig to promote the forward reaction.

Since this reaction is also exothermic in nature, it also has an equilibrium constant (K_e), which dramatically decreases with temperature, providing a true paradox. We want the reaction rate to be high, but when we raise the temperature to do this, we reduce the amount of ammonia we make. At the same time, we want high pressure to shift the reaction in favor of the product with fewer gas moles than the feed gases. High pressure equipment costs money. A typical ammonia process runs at several hundred degrees Celsius and at least 1000 psig to deal with this contradiction. Reducing the temperature and pressure of this reaction is one of the "holy grails" of chemical engineering. This same challenge has been with us for over 100 years, and along with the safety aspects of such a process is the subject of large annual meetings of chemical engineers and the companies that design and run ammonia plants.

Gas Flows

One of the more common ways of measuring gas flows is with a simple pitot tube, shown in Figure 7.17.

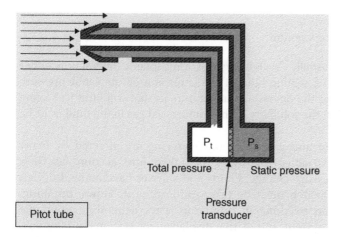

Figure 7.17 Pitot tube meter for gas flow. Source: Mendel, https://commons.wikimedia.org/wiki/File:Pitot_tube_diagram.png. Used under CC BY-SA 3.0, https://creativecommons.org/licenses/by-sa/3.0/deed.en. © Wikipedia.

These are seen occasionally on airplane wings as they are typically used to measure airplane speed.

The other measurement devices discussed earlier can also be used.

Gas Compression

Gases are normally moved by compressors (at high pressure) and fans (at low pressure). As with liquids, all the sources of pressure drop must be considered in sizing the compressor or fan. Gas flow rates are more complicated to deal with as their density will change with pressure. This requires an integrative type of flow calculations based on the changes in pressure and temperatures.

A gas compressor's work will dramatically increase the temperature of the gas (remember the differences in heat capacities between liquids and gases?), in turn affecting the efficiency of the compressor. The thermal conductivity of the gas being compressed will also change as the temperature changes. Though more capital intensive, compressing a gas isothermally (meaning cooling as the gas is compressed) will consume less energy.

As with liquid flow systems, we must take into account the following factors:

- Average versus peak flow.
- Pressure drop—have all items been considered?
- Source pressure.
- Piping design, elbows, tees.
- Capability of causing vacuum.

Compressors are usually "sealed" against atmospheric interaction either mechanically or with a sealing fluid. If the latter is used, its vapor pressure must be considered at the operating temperature as that will limit its sealing capability. In addition the solubility of the compressed gas in the fluid must be considered.

In summary, liquid and gas flows and metering are affected by many physical property variables that must be taken into account as their temperature and pressure change. Fluids and slurries have many different responses to shear, which must be taken into account. Flows, pumping, and compression can be done in numerous ways depending upon the requirements of the system. In positive displacement systems, it is important to consider the possibility of creating vacuum and the possible safety consequences.

Fluid Flow and Coffee Brewing

Let's go back to our case study again. When are fluids moving, when might their properties change, and what might be the impact? There are several different ways of brewing, but let's start with the most often used process, drip filtration. In this case, hot water is dispersed over the top of ground coffee particles, usually driven by a small pump not seen by the user, and the water flows through "extracting" or "leaching" (more on these unit operations later) flavor ingredients from the ground coffee beans and then dripping into a container. The majority of the ground bean raw material remains behind, trapped by a filter (more on this unit operation as well later) of some sort but usually a paper element. It can also be a permanent filament wound filter, expected to be washed and reused. What are the fluid issues here?

(Continued)

(Continued)

We discussed the significant impact of temperature on liquid viscosity. Is the temperature produced by the machine constant? If not, then its viscosity will be different every time a cup of coffee is brewed. If the viscosity is higher or lower, the flow rate through the ground coffee and the filter medium will also change, affecting the time that the water has to leach the flavor ingredients. The coffee will be weaker or stronger depending on the viscosity. It is unlikely that there will be any significant density differences, so this aspect can probably be ignored.

What is the nature of the pump feeding water into the coffee machine? Most likely a small positive displacement pump that "precisely" delivers the amount of water needed, but how do we know unless we check?

Water, as with any other fluid, has a surface tension, a property also affected by dissolved salts and temperature. How long does it "cling" to the coffee ground particles? How big are these particles? Are they all the same size? What is their particle size distribution? Is the water softened, adding sodium and removing calcium? Is it in the middle of a regeneration cycle? How does this affect the surface tension?

Coffee brewing is beginning to sound a bit complicated, isn't it? And we still have a long way to go!

Discussion Questions

1 What is the nature of the fluids in your process? Have all of their physical properties been measured? Is the information readily available?

2 Do you have this data over the *entire* range of actual operating conditions?

3 How were the pumps you use chosen? Is there any basis now for revisiting that decision?

4 How close to the minimum NPSH for your centrifugal pumps are you running? What would happen if you reached this point? Has that been discussed in a HAZOP or any other type of safety or process review?

5 Is there a new centrifugal pump available that would satisfy your NPSH and flow demands, should it be installed? Why or why not?

6 If you are using positive displacement pumps, what would be the consequences of the pump being "dead headed?" How long would this situation need to exist before a fluid decomposition occurred? A reactive chemical incident?

7 What types of flow meters are being used? What was the basis for choosing them originally? Has anything changed that would warrant a change? (Accuracy needed, type of fluid, pressure drop, cost?)

8 Is volumetric flow rate sufficient or would mass flow rate be more useful information? For what reason?

Review Questions (Answers in Appendix with Explanations)

1 Total fluid pressure is measured by the sum of:
A __Static pressure and dynamic pressure
B __Dynamic pressure and fluid density
C __Static pressure and anticipated friction loss
D __All of the above

2 Laminar flow implies:
A __Pressure drop for the fluid flow is high
B __Fluid is wandering around with no direction
C __Pipes are made from plastic laminates
D __Little or no mixing across the cross-sectional area of a pipe

3 Turbulent flow implies:
A __Fluid is well mixed across the cross-sectional area of the pipe
B __Pressure drop will be higher than laminar flow
C __There is little or no adhesion between the fluid and the piping wall
D __All of the above

4 Turbulent versus laminar flow will affect all but:
A __Pressure drop in the pipeline
B __Mixing in the pipe
C __Cost of the piping materials
D __Pressure drop across valves and instrumentation

5 Key fluid properties affecting fluid handling systems include all except:
A __Density
B __Viscosity
C __Residence time in tank prior to pumping
D __Temperature and vapor pressure

6 Viscosity characterizes a fluid's resistance to:

A __Being pumped

B __Being held in storage

C __Price change

D __Shear

7 The viscosity of a fluid is most affected by:

A __Density

B __Pressure

C __Index of refraction

D __Temperature

8 The viscosity of an ideal (Newtonian) fluid reacts to a change in shear at constant temperature by:

A __Remaining the same

B __Increasing

C __Decreasing

D __Need more information to answer

9 A dilatant fluid's viscosity _____ with shear:

A __Increases

B __Decreases

C __Stays the same

D __Depends on what kind of shear

10 A thixotropic fluid responds to shear by _____ its viscosity:

A __Increasing

B __Decreasing

C __Not affecting

D __Depending on what kind of shear

11 In general, adding solids to a liquid (converting it into a slurry) will _____ its viscosity:

A __Increase

B __Decrease

C __Not affect

D __Can't be known

12 The Reynolds number is:

A __Dimensionless

B __A measure of turbulence in flow

C __The ratio of diameter × density × velocity

D __All of the above

13 A dimensionless number in chemical engineering:
    A __Provides a simple way of characterizing an aspect of design
    B __Has no units (if calculated correctly)
    C __Allows a chemical engineer to estimate relative behavior of an engineering system
    D __All of the above

14 Friction in fluid flow is influenced by all but:
    A __Fluid properties
    B __Flow rate
    C __Piping design characteristics
    D __Cost of energy to pump the fluid

15 Pressure drop in a fluid system can be affected by:
    A __Length of piping
    B __Number and nature of connections and valves
    C __Degree of corrosion on walls
    D __All of the above

16 The difference between a centrifugal and positive displacement pumps is:
    A __Centrifugal pumps are less expensive
    B __Positive displacement pumps have a characteristic curve
    C __Centrifugal pumps generate constant pressure; positive displacement pumps put out constant flow
    D __It is harder to "dead head" a positive displacement pump versus a centrifugal pump

17 Centrifugal pumps require a minimum net positive suction head (NPSH) to operate; otherwise they will cavitate. This can be caused by all but:
    A __Improper placement of the pump on an engineering drawing
    B __Reducing the level of the liquid feeding the pump
    C __Raising the level of a tank into which the pump is discharging
    D __Raising the temperature of the feed liquid and raising its vapor pressure

18 If the process needs to exceed the minimum NPSH available, what options are available?
    A __Raise the height of inlet stream to the pump
    B __Lower the temperature of the inlet feed
    C __Increase the size of the piping in the system
    D __Any of the above

19 The choice of a flow meter will depend upon:

    A __Accuracy required

    B __Pressure drop tolerance

    C __Cleanliness of fluid

    D __All of the above

Additional Resources

Collins, D. (2012) "Choosing Process Vacuum Pumps" *Chemical Engineering Progress*, 8, pp. 65–72.

Corbo, M. (2002) "Preventing Pulsation Problems in Piping Systems" *Chemical Engineering Progress*, 2, pp. 22–31.

Fernandez, K.; Pyzdrowski, B.; Schiller, D. and Smith, M. (2002) "Understanding the Basics of Centrifugal Pump Selection" *Chemical Engineering Progress*, 5, pp. 52–56.

James, A. and Greene, L. (2005) "Pumping System Head Estimation" *Chemical Engineering Progress*, 2, pp. 40–48.

Kelley, J.H. (2010) "Understanding the Fundamentals of Centrifugal Pumps" *Chemical Engineering Progress*, 10, pp. 22–28.

Kernan, D. (2011) "Learn How to Effectively Operate and Maintain Pumps" *Chemical Engineering Progress*, 12, pp. 26–31.

Livelli, G. (2013) "Selecting Flowmeters to Minimize Energy Costs" *Chemical Engineering Progress*, 5, pp. 35–39.

https://commons.wikimedia.org/wiki/File:Viscosity_video_science_museum.ogv (accessed on August 30, 2016).

8

Heat Transfer and Heat Exchangers

A second major subject for chemical engineering practice is the study of energy transfer, most often the use of equipment to transfer heat, either heating or cooling. We use the term refrigeration when we are discussing cooling below ambient temperatures. We may want to heat or cool for a variety of reasons:

1) Preheat a material entering a chemical reaction system.
2) Cool a process vessel or reaction system that is generating heat in an exothermic reaction.
3) Heat a reaction to sustain its reaction if it is an endothermic reaction.
4) Condense vapors from an evaporator or distillation column.
5) Reboil liquid at the bottom of a distillation column.
6) Refrigerate a material that might decompose at a very low temperature in a reaction system. When we are cooling below room temperature, the heat exchanger is often referred to as a "chiller."
7) Collect or remove radiative heat. Solar collectors are examples of such devices. High temperature cracking furnaces are another.

The equipment and fluids used in this unit operation vary widely according to the requirements and the resource availability. For example, if it is necessary to cool a material in South Texas to 70°F in the summer, this will not be possible with lake or river water. Some kind of refrigeration will be required. If this same system were in Minnesota or the upper peninsula of Michigan, there would be no such need. Conversely, if there is a need to heat a material, the pressure of the steam available will put limits on what is possible as steam temperature and pressure go hand in hand.

As in the other topics we have covered, there are two fundamental principles that we need to remember:

1) Heat flows from high temperature to low temperature.
2) Heating or cooling can be used to increase or decrease the temperature of a material or to change its phase, that is, melt/freeze or boil/condense. When a material is undergoing a phase change, there is no change in temperature.

Chemical Engineering for Non-Chemical Engineers, First Edition. Jack Hipple.
© 2017 American Institute of Chemical Engineers, Inc. Published 2017 by John Wiley & Sons, Inc.

There are also two different types of heat transfer, convective and conductive. *Convective* heat transfer involves bulk transfer of heat *between* materials. Mixing of hot and cold fluids together or the heating or cooling of a mass of fluid are such examples. *Conductive* heat transfer involves the flow of heat *through* a material with no temperature change. An example here is the loss of heat from the interior of a house to the outside. Even though there is no apparent bulk movement of the materials (say, in a rod or in the insulation inside the walls of the house), the molecules inside these materials are vibrating and moving, transporting the energy from hot to cold.

The conductivity of materials is measured by a property known as the thermal conductivity or k, whose units are BTU/h/ft/°F or cal/s/m/°C or some other combination of heat, time, and distance that are consistent. The higher a material's k value, the more rapid thermal energy (heating or cooling) moves through it. If we want to prevent heat flow, we choose a material with a low k; if we want to maximize the rate at which heating or cooling leaves a system, we choose a material with a high k.

The thermal conductivity values for some materials are shown in Table 8.1. You can think about these numbers in the context of cooking and the pots and pans we use every day. Many of the more expensive (since copper is more expensive than steel or aluminum) pans are copper plated or made entirely of copper to provide quick, even heat. Aluminum is a compromise between the low k of steel and the high k of copper as well as the prices of each of these. It would be wonderful if we could afford to make cooking pans made out of silver (or process heat exchangers!). As you can see, brick has a very low k, making it an insulating material.

Table 8.2 shows some additional k values that provide some interesting insights. Notice the large difference between a hydrocarbon, such as hexane, and water. The use of water in a heat exchanger would require significantly less material to transfer the same amount of heat. Note that the gases have relatively low k's, meaning that it would take less heat input to raise their temperature versus solids and liquids. Notice also the extremely large k value for sodium

Table 8.1 Thermal conductivities of common materials.

Aluminum	115–120 (BTU/h ft °F)
Copper	215–225
Iron/steel	25–40
Silver	240
Brick	0.08–0.12

Source: Average of public sources.

Table 8.2 Thermal conductivities of materials.

N-Hexane	0.08 (BTU/h ft^2 °F)
Water	0.36
Sodium metal	45
Hydrogen	0.11
Methane	0.02
Air	0.016
Argon	0.010

Source: Average of public information.

metal, meaning that a small amount of sodium can absorb tremendous quantities of heat. This is the primary reason that liquid sodium is used as an emergency coolant in nuclear power plant reactors.

In most storm windows, there is an insulating layer of air between the glass panes to inhibit the loss of heat. In far northern climates, it can pay to inject and seal argon (an expensive gas) between the panes of glass to further inhibit heat flow. Its *k* is 40% less than air.

When we have bulk material flow and heat transfer is involved, we have *convective* heat transfer. This can be, as mentioned before, the bulk mixing of a hot and cold fluid, but in the chemical engineering world, it is usually referring to two fluids (liquids or gases) being heated or cooled with a metal barrier between them in the form of a *heat exchanger*.

Types of Heat Exchangers

Though not an all-inclusive list, the following are the primary types of heat exchange equipment in use in the chemical industry:

1) Shell and tube exchangers. These types of exchangers are the workhorse of the chemical industry. An illustration of such an exchanger, in a typical countercurrent flow design, is shown in Figure 8.1.

   In this type of equipment, a cooling fluid outside the tubes is being used to cool a hot process stream inside the tubes. The flow inside the tubes and outside the tubes is where the convective heat transfer is occurring. There is conductive heat transfer across the metal walls of the tubes. Though this illustration shows a "one-pass" configuration, the tubes can be of a "U tube" configuration and have a multi-pass configuration. The baffles are added to increase the turbulence of the fluid on the shell side of the exchanger, increasing the heat transfer rate.

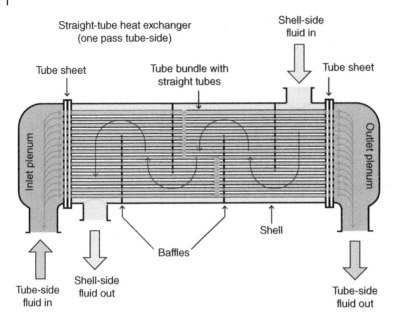

Figure 8.1 Conventional shell and tube heat exchanger. Source: Pedlackas, https://commons.wikimedia.org/wiki/File:Straight-tube_heat_exchanger_1-pass.PNG.

2) Jacketed Pipe Exchangers (see Figure 8.2).

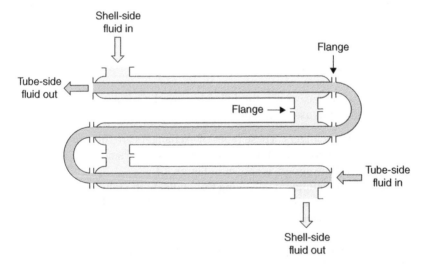

Figure 8.2 Jacketed pipe heat exchanger. Source: https://commons.wikimedia.org/wiki/File:Double-Pipe_Heat_Exchanger.png. Used under CC BY-SA 4.0, https://creativecommons.org/licenses/by-sa/4.0/. © Wikipedia.

These are simply pipes with jackets wrapped around them (similar to insulation) that either cool or heat. Due to the area and time limitations, these can be used only for minimal amounts of heat transfer.

3) Chillers. This terminology refers to any kind of heat exchanger operating below normal ambient temperatures. Usually its purpose is to condense low boiling point compounds and use a refrigerant such as ammonia, low boiling hydrocarbon, or a fluorinated hydrocarbon. It could refer to any number of mechanical configurations.

4) Condensers. This refers to a heat exchanger being used to condense a vapor stream into a liquid. It could refer to any number of mechanical configurations. It is most frequently seen used to condense the overhead vapor from a distillation column (distillation will be discussed in detail in Chapter 10) as seen in Figure 8.3.

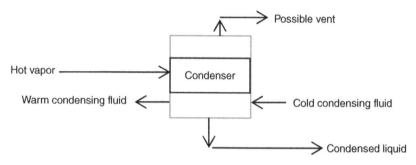

Figure 8.3 Condenser.

This type of heat exchanger can also be used to condense and remove volatile organic carbons (VOCs) from a gaseous vent stream. Depending upon the composition of the incoming stream, not all of the incoming vapor will condense, and a vent may be necessary and further handling of the vent stream required. If only part of the stream is condensed, this type of condenser would be referred to as a partial condenser.

5) Reboilers. This refers to a heat exchanger (again, could be a number of mechanical configurations) that is being used to "reboil" liquid at the bottom of a distillation column (more later on this unit operation in Chapter 10) as shown in Figure 8.4.

The overall heat transfer coefficient of a heat exchanger is normally referred to as U. Given the basic equation of $Q = UA\Delta T$, the heat transfer area required is simply $A = Q/U\Delta T$. The Q is simply the amount of energy required and can be calculated by multiplying the fluid rate times its heat capacity times the temperature change (and latent heat of vaporization if applicable) and then dividing by the estimated heat transfer coefficient.

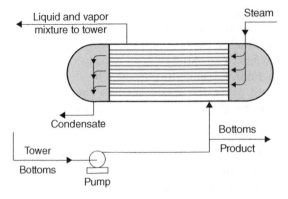

Figure 8.4 Reboiler. Source: Baychok, https://commons.wikimedia.org/wiki/
File:ForcedCirculation.png. Used under CC BY-SA 3.0 https://creativecommons.org/licenses/
by-sa/3.0/deed.en. © Elesvier.

Countercurrent versus Cocurrent Flow Design

1) It is possible to configure a heat exchanger in two different ways usually described as cocurrent (sometimes referred to as parallel flow) and counter-current (sometimes referred to as counterflow). An example of a typical countercurrent exchanger was shown in Figure 8.1. In cocurrent flow, both hot and colder streams enter at the same end of the heat exchanger as shown in Figure 8.5.

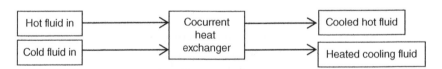

Figure 8.5 Flows in a cocurrent heat exchanger.

2) The difference in profiles of the fluid temperatures will be quite different, since with cocurrent flow, it is not possible for either the hot or cold fluid to achieve more than their "average" temperature. The temperature profiles within these two different types of exchangers would be as shown in Figure 8.6.
3) A countercurrent flow arrangement will always allow for a colder exiting temperature of the fluid being cooled.

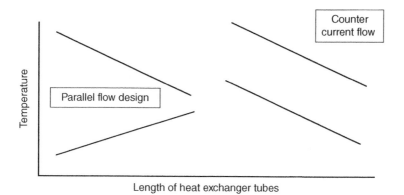

Figure 8.6 Cocurrent versus countercurrent temperature profiles.

Heat Transfer Coefficient

The overall heat transfer coefficient, U in $BTU/h\,ft^2$ or similar metric units, is a sum of three different heat transfer coefficients:

1) Convective heat transfer from the cooling fluid to the wall
2) Conductive heat transfer through the wall
3) Convective heat transfer from the inner cooling wall surface to the process fluid

The overall heat transfer coefficient, U, is the sum of these three individual heat transfer coefficients, $h_{outerwall} + h_{tube} + h_{innerwall}$. In many situations where the Reynolds number is high (turbulence is high) on both the inner tube and outer tube, the h across the tube is the highest number, meaning that it provides the least resistance to heat flow. The normal challenge is getting the heat to and from the wall. However, this should not be assumed if the fluids have a low Reynolds number. The individual h's can be calculated (see Table 8.3), and the h of the tube calculated from its thermal conductivity, thickness, and the temperature difference across the tube. These rough calculations will tell us whether the h across the tube is small enough to be ignored.

The amount of energy transferred is equal to the overall heat transfer coefficient (actually a sum of three sub-coefficients) times the area of the tubing times the temperature differential at any point.

Approximate ranges for these overall heat transfer coefficients are shown in Table 8.3.

Table 8.3 Approximate heat transfer coefficients.

Outside fluid	Inside fluid	Contact material	Overall U
Air	Water	Metal	1.5–2.5
Water	Water	Metal	40–80
Steam	Water	Metal	120–180

Source: Reproduced with permission of Engineering Toolbox.

Heat transfer coefficients are a function of a number of fluid physical properties as well as flow conditions. One of the equations used to calculate these in more detail uses a grouping of dimensionless numbers (we have already been introduced to the Reynolds number, which is a measure of turbulence of flow) along with the geometry of the piping used in the heat exchange equipment:

$$\frac{hD}{k} = 0.023\left(\frac{DV\rho}{\mu}\right)^{0.8}\left(\frac{C_p\mu}{k}\right)^{0.3}$$

where h is the heat transfer coefficient we have been referring to, D is the pipe diameter, k is the thermal conductivity of the fluid, $DV\rho/\mu$ is the Reynolds number, and $C_p\mu/k$ is a dimensionless number that we know as the Prandtl number, which is a measure of the physical properties of the material, independent of whether the fluid or gas is flowing. It is basically the ratio of the fluid's ability to absorb heat (C_p) versus transfer heat (k) and the rate at which heat can be moved, as indirectly indicated by its viscosity. One only needs to think about the difference in moving heat through a low-viscosity fluid such as water versus a high-viscosity fluid such as maple syrup. If we rearrange this equation to solve for h, we see a relationship like the following:

$$h = \frac{0.023\left(DV\rho/\mu\right)^{0.8}\left(C_p\mu/k\right)^{0.3}(k)}{D}$$

Without worrying about the exact numbers, we can look at the variables and how this equation predicts how the heat transfer coefficient would respond to a change in conditions or physical properties:

1) If the Reynolds number increases, the heat transfer will *increase* by the 0.8 power. The greater the turbulence, the more efficient the heat transfer, but it does not increase linearly.
2) If the pipe diameter increases, the heat transfer will *decrease* by the 0.2 power (the smaller the pipe diameter, the lower the velocity and less turbulent is the flow).

3) If the viscosity increases, the heat transfer will decrease by the 0.5 power (if the fluid is "thicker," the Reynolds number decreases and the ability to mix the fluid drops). Again, think about water versus maple syrup.
4) If the thermal conductivity of the fluid increases, the heat transfer will increase by the 0.7 power.

The point here is not to memorize the equation, but just to reinforce the natural logic that heat transfer will change in response to variables that can be affected and changed in the chemical engineering design of not only the heat exchanger but also the choice of fluids and their physical properties. This kind of general knowledge can also assist in estimating changes in heat transfer efficiency when changes to process systems are made.

Utility Fluids

It is easy to overfocus on the process stream that needs to be heated or cooled, but the utility fluid (water, steam, refrigerant, or heat transfer fluid) is equally important. Their properties and availability may not be under complete control of the user, especially if they are supplied by a public utility. Even though we may specify 150 psig steam on a process flow sheet as well as on our calculations, the chances of the steam being at exactly 150 psig at any given time are slim to none, despite the best intentions of the utility manager or supplier. What happens if it's 140 psig? 160 psig? Can the process fluid overheat? Are there any consequences to this? (Recall our discussions about HAZOP thinking earlier.) If the process fluid is not heated sufficiently, does the reactor it may be feeding slow down in reaction rate? Is this a problem? If so, what kind of problem? How do you counteract? What happens if the steam supply is totally lost?

If the utility fluid is water (used as a cooling fluid), what are the consequences if its temperature is too hot or too cold? Is it cooling an exothermic reaction? Could there be a runaway reactive chemicals incident? How many layers of protection are needed? What is the backup plan, especially if the cooling water is coming from a public utility over which you have no direct control? How do you decide? Is the output from this "cooled" reactor feeding some other process? What are the consequences of losing this flow?

In a chiller, low boiling compounds such as hydrocarbons, fluorinated hydrocarbons, and liquid ammonia are used as coolants. Since these materials, when vaporized, generate pressure, it is critical to make sure that valves within the heat exchanger system are not closed inappropriately, potentially causing the rupture of a heat exchanger that was not designed to be a pressure vessel.

Air Coolers

If there is a shortage of cold cooling water, air coolers are frequently used:

Figure 8.7 Air-cooled heat exchanger. Source: Reproduced with permission of Engineering Toolbox.

These types of heat exchangers have much lower heat transfer coefficients than those using fluids, but they use a "free" resource. Ambient air has its limits in terms of temperature, so it is important to consider the worst-case air temperature in designing the required area.

Scraped Wall Exchangers

When the materials that need to be heated or cooled have high viscosities, the tube fouling may be so severe that continuous scraping and clearing of the walls may be required, as seen in the illustration of a commercial scraped wall exchanger in Figure 8.8.

Figure 8.8 Scraped wall exchanger. Source: Reproduced with permission of Sulzer. © Elsevier.

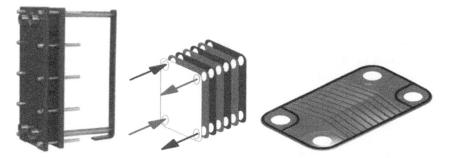

Figure 8.9 Plate and frame heat exchanger. Source: Varem, https://commons.wikimedia.org/wiki/File:Plate_heat_exchanger.png. © Wikipedia.

Plate and Frame Heat Exchangers

These are shell and tube exchangers with very tight clearance between sides of the exchanger and high turbulence between the layers (Figure 8.9). They typically have high heat transfer efficiencies but, due to close clearances, are not the first choices for slurries. Their major advantage is high heat transfer rates in a small area.

Leaks

Any heat exchanger will eventually leak between the shell and tube side. It's just a question of when. The important issue of concern is when this happens which way do we want the exchanger to leak? Shell to tube, or tube to shell? This will be primarily controlled by the pressure values on either side, which can be considered in the design. If several tubes break off the tube sheet (see section "Mechanical Design Concerns"), massive leaks are possible. What kind of safety, shutdown, and environmental protections need to be in place not only for the heat exchanger but also for its downstream flow, and its use?

Mechanical Design Concerns

Though the detailed mechanical design, including welding and meeting vessel codes, is typically done by mechanical and welding engineers, there a few basic points to remember when operating an exchanger, especially one that undergoes a large temperature change:

1) Thermal Expansion. Metals have coefficients of expansion and will stretch or contract as their temperature rises or falls. Since metals are typically

joined by welding, too rapid a change in temperature may produce sufficient force to rupture the weld.

2) Fouling. Heat exchangers will foul or develop coatings on the tube and shell surfaces over time due to a variety of reasons including contamination, impurities in flow streams, and corrosion products from the reactions of the process and utility fluids with the materials in the heat exchanger.

3) Placement of Fluids. There may be specific reasons we might want the process or utility fluid to be either on the shell or tube side, but there are also some general suggestions that are helpful:
   a) Pressure drop will normally be lower on the shell side, so if flow pressure has some limitations, this may be a consideration.
   b) Cleaning the tubes is easier than taking the assembly apart, so if fouling is expected, it would be better to put that fluid on the tube side.

4) Chiller Pressure Development. If a refrigerant such as ammonia is being used, we must remember that the cold liquid can vaporize and generate pressure, so it is critical in maintenance procedures to remember that the heat exchanger can turn into a pressurized vessel, which it was not designed to be.

5) Leaks. At some time point in time, an exchanger will leak between the shell and the tube sides. It is not a matter of if; it is when. This requires that we think about the consequences of this event. Which way do we want the fluid to leak? Tube into shell? Or shell into tube? What are the consequences of these choices from the standpoint of reactive chemicals, product contamination, or environmental release?

Cleaning Heat Exchangers

All heat exchangers, regardless of type of design or service, will need cleaning. There are a number of ways that this is done. First, mechanically. Tubes can be "rodded" out mechanically and the waste disposed of properly. Second, ultrahigh pressure water jet cleaning can be used. It must be shown that the high pressure will not cause some other rupture in the equipment. It is also important to provide education on the danger of high pressure water jets, which can amputate fingers and toes. Third, chemical cleaning methods can be used. There are inhibited acid solutions available that, for a period of time, will dissolve the corrosion product without corroding the base steel metal. In addition to ensuring that the proper conditions are maintained, the proper safety equipment must be worn while handling these acid streams.

Radiation Heat Transfer

Every "thing" or person emits radiative energy, but at room temperatures, the amount of this energy is extremely small and is not considered in normal heat transfer calculations. But we all are concerned about sunburn in the summer time from a star emitting radiation over 90 million miles away. Even at this distance the extremely high temperature makes this a serious heat transfer issue.

In the real world, temperatures need to exceed 1000°C for radiation heat transfer to be of a significant amount that needs to be considered. This does occur in hydrocarbon cracking furnaces and in carbon black plants. In high temperature hydrocarbon cracking furnaces, high molecular weight hydrocarbons are "cracked" into lighter, more valuable materials such as gasoline, ethylene, propane, diesel fuel, and aromatics. Side issues with these types of furnaces include the need for high temperature insulation materials and the carburization of tubes requiring eventual cleaning.

High Temperature Transfer Fluids

In some processes it is desired to heat materials to 300–600°C. If this is done with steam, extremely high pressure steam is required, normally beyond the capability of conventional boilers. A class of organic compounds, with low vapor pressure at high temperatures, has been developed over the years by a number of manufacturers to meet this need. These compounds are typically organic fluids with very low boiling points and low viscosities, allowing them to be used to supply heat input at very high temperature at significantly less pressure than the same temperature supplied by steam. Many of these fluids are mineral oils or diphenyl oxides, but many have proprietary formulations. Since these fluids are typically organic liquids, the addition of a fire hazard must be considered. Figure 8.10 shows the range of usability of these fluids from one manufacturer.

At one extreme, the fluid will freeze, and at the other extreme, it will decompose rapidly. The chemical synthesis and composition of these various fluids are controlled by their manufacturers to meet specific customer needs.

Since these fluids are organic, they will decompose slowly over time via oxidation or thermal decomposition and are monitored by the customer and supplier and, at some point in time, will have to be disposed of and replaced. Decomposition of these fluids takes the form of caking inside the heat transfer piping as shown in Figure 8.11.

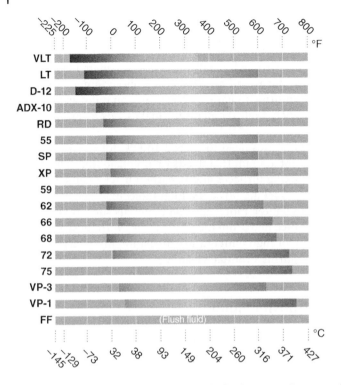

Figure 8.10 Thermal ranges for heat transfer fluids. Source: Courtesy of Eastman Chemical Company. © Elsevier.

Figure 8.11 Pipe plugged with decomposed heat transfer fluid.

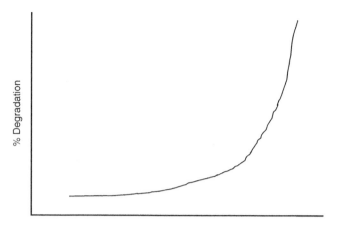

Figure 8.12 Heat transfer fluid degradation time and temperature.

In practice, the provider of these fluids works with the user to develop a systematic sampling process to measure the degradation rate, and at an appropriate time, the fluid will be flushed and replaced (Figure 8.12).

Another key aspect of virtually all of these fluids is that, unlike steam, they are flammable to varying degrees. This requires users of these fluids to have on-site firefighting and fire safety as part of their overall safety programs.

Summary

Heat transfer and the equipment required to transfer heat are integral parts of a chemical process and a chemical complex. There are numerous types of equipment that operate under basic heat transfer equations but have advantages and disadvantages depending upon the nature of the fluid requiring heating or cooling, the utility fluids available, the consideration for radiative heat transfer, and the necessity for equipment cleaning.

Coffee Brewing: Heat Transfer

Back to our coffee brewing with our chemical engineering hats on. Do we want coffee brewed with cold water? Maybe REALLY cold water if we want iced coffee, but let's stay focused on normal hot coffee. We need to heat the water. How do we do this? By using the $U = QA\Delta T$ equation. What is the U? Well, how much water do we want to heat? What is the starting temperature? How hot do we want it to be? That's the ΔT. How much water? We decide this

---

(Continued)

(Continued)

when we decide how many cups we are brewing. What's the *A*? What is the heat transfer area inside the coffee pot? We don't really know, but someone who worked for the coffee brewing machine manufacturing company decided that.

What about the "hot plate?" It has a surface area and the manufacturer has decided on a temperature to be sustained. We will discuss evaporators later, but that's what a hot plate is; while it is maintaining a temperature, some of the water is evaporating, concentrating the coffee, AND supplying heat input that causes the coffee flavor ingredients to degrade into not so tasty materials including aldehydes. This is no different than the kinetic rate material we discussed earlier. What is another alternative that we have seen in home coffee brewers? A sealed vacuum receiver without a hot plate. While there is still some degradation going on, its rate is lower since there is not constant heat input.

And your coffee cup? It can be a simple cup (which will lose heat at the $Q = UA \Delta T$ rate). What affects this? What is the area of the cup? What is the temperature difference between the coffee and the room? What is the thermal resistance across the wall of the cup? Was the initial temperature affected by the addition of cold cream? How much? Sugar? How much? We have seen practical attempts to minimize this by insulated cup holders and small hot plates in an office. Keep thinking about this process!

Discussion Questions

1 What is the heat balance around your process or area you are involved in? Is it calculated by an online process control computer? If so, what aspects of the operation could change that may cause errors in the calculation?

2 Do you have thermal property data over the entire range of your process? What situations could cause the process to go outside this range? Are the thermal properties taken into account?

3 If the temperature of your feed into a reactor increases, what happens?

4 If the temperature of the feed decreases, what happens?

5 If you are using high temperature heat transfer fluids, do you know their decomposition point? How often are they sampled? What happens if their process lines totally plug? Do you have adequate fire protection? On-site? Off-site?

6   If you have a shell and tube heat exchanger, which direction will it leak? Shell to tube? Tube to shell? What happens when either fluid contaminates the other? Safety issues? Quality issues? Reactive chemicals issues?

7   If you are using a refrigerant or some other fluid under pressure on one side of the exchanger, what happens if the outlet flow of the liquid under pressure is blocked? Is the exchanger designed to handle the pressure that results? What does the energy balance look like in this situation? How much heat is entering? How fast does the other side of the exchanger heat up? To what pressure? Is there an environmental issue?

8   How did you decide on the heat exchanger equipment and design that you now have? Has anything changed that would warrant a technical review?

9   What is the source of your utility fluid? Is it reliable? What happens if its flow stops? How does its temperature and pressure affect the operation of your heat exchangers?

Review Questions (Answers in Appendix with Explanations)

1   An energy balance around a process or piece of equipment requires knowledge of all but:
    A __Flow rates and temperatures of flows in and out
    B __The speed of the pump feeding the vessel
    C __Heat generated by any reaction occurring
    D __Heat capacities of streams in and out

2   The three methods of heat transfer are all but:
    A __Conductive
    B __Convective
    C __Convoluted
    D __Radiation

3   Conductive heat transfer refers to heat moving:
    A __Above
    B __Below
    C __Around
    D __Through

4   Convective heat transfer refers to heat moving:
   A __In a bulk fashion
   B __Only as a function of convective currents
   C __On its own
   D __None of the above

5   Variables that affect the rate of heat transfer are:
   A __Flow rates
   B __Physical properties of fluids
   C __Turbulence within the heat transfer area or volume
   D __All of the above

6   If the pipe diameter is increased and all other variables remain the same, the rate of heat transfer:
   A __Will increase
   B __Will decrease
   C __Will stay the same
   D __Can't tell without more information

7   If the viscosity of fluids on the shell side is increased, the heat transfer rate:
   A __Will increase
   B __Will decrease
   C __Will stay the same
   D __Can't tell without more information

8   The utility fluid is:
   A __The fluid that costs less since it is asked to do anything
   B __A fluid that can move in either direction
   C __The non-process fluid in a heat exchanger
   D __A utility that is a fluid

9   The overall heat transfer coefficient includes the resistance to heat transfer through:
   A __The pipe wall
   B __The barrier layer on the shell side
   C __The barrier layer on the tube side
   D __All of the above

10   Design issues with heat exchangers include all but:
   A __Area required
   B __Corrosion resistance to the fluids
   C __Leakage possibilities
   D __The dollar to euro conversion at the time of design

11 The primary design limitation of air cooled heat exchangers is:
A __Fan speed
B __Distance from a river or lake
C __Temperature of outside air
D __Contractor's ability to raise or lower the heat exchanger

12 Fouling and scaling on a heat exchanger can be caused by:
A __Deposition of hard water salts
B __Softness of the heat exchanger material
C __Use of distilled water as a coolant
D __Poor maintenance

13 Radiative heat transfer can be an important concern in:
A __Sunburns while working in a chemical plant in Houston
B __Chemicals that are red or yellow
C __Insufficient heat transfer on a cloudy day
D __High temperature processing in the oil and petrochemical industry

14 High temperature heat transfer fluids are used when:
A __Cold ones are not available
B __It is necessary to transfer heat at high temperature and low pressure
C __Hot water is not available
D __The plant manager owns stock in a company that makes and sells them

15 The downside of high temperature heat transfer fluids include all but:
A __Flammability
B __Possible degradation and need to recharge
C __Potential chemical exposure to the process fluid
D __Ability to transfer high temperature heat at low pressure

Additional Resources

Arsenault, G. (2008) "Safe Handling of Heat Transfer Fluids" *Chemical Engineering Progress*, 3, pp. 42–47.

Beain, A.; Heidari, J. and Gamble, C. (2001) "Properly Clean Out Your Organic Heat Transfer System" *Chemical Engineering Progress*, 5, pp. 74–77.

Bennett, C.; Kistler, R. and Lestina, T. (2007) "Improving Heat Exchanger Designs" *Chemical Engineering Progress*, 4, pp. 40–45.

Beteta, O. (2015) "Cool Down with Liquid Nitrogen" *Chemical Engineering Progress*, 9, pp. 30–35.

Bouhairie, S. (2012) "Selecting Baffles for Shell and Tube Heat Exchangers" *Chemical Engineering Progress*, 2, pp. 27–32.

Chu, C. (2005) "Improved Heat Transfer Predictions for Air Cooled Heat Exchangers" *Chemical Engineering Progress*, 11, pp. 46–48.

Gamble, C. (2006) "Cost Management in Heat Transfer Fluid Systems" *Chemical Engineering Progress*, 7, pp. 22–26.

Garvin, J. (2002) "Determine Liquid Specific Heat for Organic Compounds" *Chemical Engineering Progress*, 5, pp. 48–50.

Krishna, K.; Rogers, W. and Mannan, M.S. (2005) "Consider Aerosol Formation When Selecting Heat Transfer Fluids" *Chemical Engineering Progress*, 7, pp. 25–28.

Laval, A. and Polley, G. (2002) "Designing Plate-and-Frame Heat Exchangers" (3 part series), *Chemical Engineering Progress*, 9, pp. 32–37; 10, pp. 48–51, and 11, pp. 46–51.

Lestina, T. (2011) "Selecting a Heat Exchanger Shell" *Chemical Engineering Progress*, 6, pp. 35–38.

Nasr, M. and Polley, G. (2002) "Should You Use Enhanced Tubes?" *Chemical Engineering Progress*, 4, pp. 44–50.

Schilling, R. (2012) "Selecting Tube Inserts for Shell and Tube Design" *Chemical Engineering Progress*, 9, pp. 19–25.

9

Reactive Chemicals Concepts

This chapter is a collection of issues relating to reactivity and control of heat-generating chemical reactions, both of which, if not properly considered and accommodated in design, can cause significant consequences, including loss of life and property. There are many inherently unstable families of chemicals (i.e., peroxides) that are well known to be thermally unstable and must be stored under refrigeration. In these cases, it is essential to provide backup utility supply to the refrigeration system. The separation of oxidizing and reducing gas cylinders is also a common practice to minimize the possibility of these types of materials interacting with each other, producing a large release of energy. Chemicals that have known degradation rates to hazardous by-products must be monitored on a regular basis, especially if in storage situations where they are not seen regularly.

The safety and operational concepts in the processing of materials that could generate uncontrolled release of energy are the most important in chemical engineering. It concerns the balance of cooling to remove energy released and an exothermic chemical reaction, mixing, or degradation process, which is heat generating. We have discussed the two key aspects within this area already—kinetics and heat transfer. One is a logarithmic function of temperature (kinetics, $k = Ae^{-E/RT}$) and the other a linear function ($Q = UA\Delta T$). Let's look at a typical rate expression as a graph as we reviewed earlier (Figure 9.1).

If we superimpose a heat transfer (a linear function) line on top of this, we see a graph such as that shown in Figure 9.2.

When the rate of energy generation from the reaction is greater than the system's ability to remove the heat, we have a runaway chemical reaction. It is absolutely critical that this point be known for any exothermic chemical reaction system and to think seriously about how to minimize the possibility of reaching this point, how to monitor and shut off feeds, how to provide supplementary cooling, or other means to minimize the consequences.

The shapes of these curves, as well as their response to temperature and other process conditions, have been covered in the previous chapters (4 and 8)

Chemical Engineering for Non-Chemical Engineers, First Edition. Jack Hipple.
© 2017 American Institute of Chemical Engineers, Inc. Published 2017 by John Wiley & Sons, Inc.

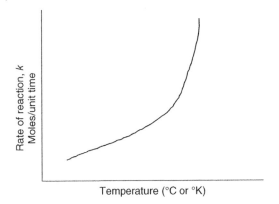

Figure 9.1 Rate of reaction versus temperature.

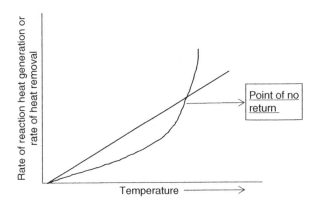

Figure 9.2 Point of no return for a runaway reaction.

on kinetics/reaction engineering and heat transfer. Reactive chemicals analysis includes a full knowledge of the nature of the chemicals being handled and the integration of kinetics and heat transfer to provide a complete view of the safety implications of running any exothermic reaction. This knowledge can also be used to design appropriate backup utility supplies as well as the control systems needed to ensure proper shutdown of feeds to an exothermic reaction.

As a suggested review on this topic, the Chemical Safety Board video on the T-2 explosion that occurred in Jacksonville, FL is recommended:

http://www.csb.gov/t2-laboratories-inc-reactive-chemical-explosion/

In this case, in addition to the unawareness of the general hazard, the scale-up of this known exothermic chemical reaction to larger and larger reaction volumes, moved the system along the curve as seen in Figure 9.2. The cooling capability was limited by both the reactor jacket heating area and the temperature of the cooling water. As the reaction scale was increased, the heat generation increased beyond the capability of the heat removal system, causing excessive pressure generation and the release of hydrogen gas (a reaction by-product).

There have been many reactive chemicals incidents in the chemical industry over the past few decades caused by one or more of the following:

1) Insufficient data on reaction rates and their change with process conditions
2) Insufficient heat transfer area to remove the heat of an exothermic chemical reaction, especially beyond the normal operating conditions
3) Lack of backup cooling water supply, especially if primary cooling is from a public utility
4) Lack of communication of KNOWN information to plant operating personnel
5) Insufficient numbers of layers of protection in the event of failure of primary means of controlling feeds, pressure, or cooling
6) Multiple charging of materials
7) Mechanical shock or extremes in cold or heat
8) Unrecognized side reactions
9) Insufficient or sudden loss of agitation or mixing
10) Friction and shock

Summary

Though one of the shortest chapters in this book, it is the most important. It is at the intersection of chemistry, chemical engineering, and mechanical engineering. The lack of complete knowledge of this topic has been the cause of many injuries, deaths, and huge property losses in the chemical industry. The topics reviewed here need to be at the heart of any comprehensive safety and loss prevention program. They are frequently included in most HAZOP and FMEA reviews, but if not, a separate and focused review is recommended.

Coffee Brewing: A Potential Reactive Chemical Process?

Since we are not running a reaction in the classical sense, the short answer to this is probably not. However if we put all of our ChE hats on, there are a few questions we might ask. If and when the water overflows, is there anything in your kitchen it could react with? Could it short out an electrical circuit? We usually put the spent grounds down the sink drain without thinking. Could something be in the sink or drain (left by someone previously) that it could react with? Most of the time, there's also a waste grinder below our sinks. It is putting in grinding energy. Where does that energy go? Could it cause a reaction through its heat generation? You may be putting cream, artificial cream, creams with flavorings, sugar, and sugar substitutes (what kind?). Any reactive chemical possibilities?

Discussion Questions

1 For any exothermic reactions, is the amount of heat release as a function of reaction temperature? How is this affected by any change in stoichiometry in the reaction?

2 Has the "runaway" or point of no return been clearly defined for the reaction?

3 What variables, in addition to stoichiometry, could affect it? Are they measured?

4 What is the emergency response plan for a possible runaway?

5 If the reaction produces a gas, where does it vent? Does this cause additional areas of concern?

6 How does a potential loss of agitation or mixing affect the rates of reaction and heat removal? How many layers of protection are appropriate?

7 Has a formal reactive chemicals review been made of the process? How long ago? What has changed?

8 If the cooling in the process is supplied by a public utility, what is the effect of its loss? What is the communication process between plant operations and the utility supplier?

9 What other utility losses could cause a reactive chemical release? Air? Electricity? What are the backup plans? How many layers of protection are needed?

10 Has all known reactive chemicals information been communicated to all personnel in the manufacturing plant?

11 For any endothermic reactions, what are the consequences of a loss of energy input if the feeds to the reaction continue?

Review Questions (Answers in Appendix with Explanations)

1 Reactive chemicals reviews start with an understanding of:
   A __How reactive management is to safety incidents
   B __A summary of last quarter's reactive chemicals incident reviews
   C __The chemical stability of all chemicals being handled
   D __The cost of changing storage conditions for gas cylinders

2 Reactive chemicals analysis would include all but the following:
   A __Management's reaction to a reactive chemicals incident
   B __Shock sensitivity
   C __Temperature sensitivity
   D __Heat generation during any processing

3 When considering the reactive chemicals potential of an exothermic chemical reaction, the key consideration is:
   A __The cost of cooling versus heating
   B __The cost of relief devices and environmental permits relating to an over-pressured reactor
   C __The rate of heat generation versus the rate of cooling required
   D __The possible rise in cost of processing

4 The reason there is a basic conflict between kinetics and heat transfer is that heat transfer is a linear function and kinetics or reaction rates are typically:
   A __Logarithmic with temperature
   B __Inversely proportional to pressure
   C __Subject to residence times in the reactor
   D __Vary with the square root of the feed ratios

5 A rise in reactor temperature will:
   A __Increase the rate of heat removal from the reactor
   B __Increase the rate of any chemical reaction occurring
   C __Lower the viscosity of any liquids in the reactor, increasing the heat transfer rate
   D __All of the above

6 An increase in volume used within a chemical reactor will have what effect on the potential for a reactive chemical incident?
   A __None
   B __Make the system less susceptible
   C __Make the system more susceptible
   D __Need additional information to answer

7 A drop in temperature within the reactor will ____ the probability of a runaway reaction.
   A __Increase
   B __Decrease
   C __Make no difference
   D __Need additional information

8 Improper storage of materials in warehouses can be a source of reactive chemicals incidents if:
   A __Moisture-sensitive materials are stored under a leaky roof
   B __Oxidizers and reducers are stored next to each other
   C __Known compound stability time limits are exceeded
   D __Any or all of the above

Additional Resources

Baybutt, P. (2015) "Consider Chemical Reactivity in Process Hazard Analysis" *Chemical Engineering Progress* 1, pp. 25–31.

Chastain, J.; Doerr, W.; Berger, S. and Lodal, P. (2005) "Avoid Chemical Reactivity Incidents in Warehouses" *Chemical Engineering Progress* 2, pp. 35–39.

Crowl, D. and Keith, J. (2013)"Characterize Reactive Chemicals with Calorimetry" *Chemical Engineering Progress* 7, pp. 26–33.

Johnson, R. and Lodal, P. (2003) "Screen Your Facilities for Chemical Reactivity Hazards" *Chemical Engineering Progress* 8, pp. 50–58.

Lieu, Y.-S.; Rogers, W. and Mannan, M. (2006) "Screening Reactive Chemicals Hazards" *Chemical Engineering Progress* 5, pp. 41–47.

Murphy, J. and Holmstrom, D. (2004) "Understanding Reactive Chemical Incidents" *Chemical Engineering Progress* 3, pp. 31–33.

Richardson, K. (2015) "Tool Helps Predict Reactivity Hazards" *Chemical Engineering Progress* 4, p. 21.

Saraf, S.; Rogers, W. and Mannan, M.S. (2004) "Classifying Reactive Chemicals" *Chemical Engineering Progress* 3, pp. 34–37.

10

Distillation

Of all the unit operations discussed, distillation is the one most unique to chemical engineering. It is the most widely used unit operation in the oil and petrochemical industries and is used in virtually any chemical process where liquids of differing compositions must be separated, solvent must be recovered, and hydrocarbon feedstocks such as crude oil and natural gas liquids must be separated into useful components for further processing.

The basic concepts of distillation are quite simple. It is a means of separating two or more liquids based upon their differences in boiling point and vapor pressure. Every liquid has a boiling point at a given pressure. For example, water boils at 100°C at atmospheric pressure. Between 0°C and 100°C, its vapor pressure increases until it equals the atmospheric pressure, at which point it boils. The temperature at which this happens will be a function of pressure. Each liquid has a vapor pressure curve that shows the relationship between the vapor pressure of a liquid and its temperature. Figure 10.1 shows vapor pressure curves for a number of different liquid compounds. "Normal" boiling points are normally referred to as the boiling point at atmospheric pressure. Pressure will affect the boiling point of any liquid.

The vapor pressure curve data for water is shown in Figure 10.2.

These graphs also illustrate a few additional physical property characteristics that exist for water and may exist with other liquids. Water has what is called a "triple point." This is a pressure and temperature at which all three phases (solid, liquid, and gas) can coexist. For water, this is around 4mm absolute pressure. The "critical point" is where the vapor pressure of the material will not allow its condensation.

Figure 10.3 shows the vapor pressure information for ethanol (C_2H_5OH) at atmospheric pressure.

Since there is a difference in vapor pressure and boiling point, it is conceivable that we could use distillation to separate these two materials. How? Since the shape of the curves is different and there is a difference in boiling point (ethanol boils at a lower temperature (78°C) than water (100°C)), we would

Chemical Engineering for Non-Chemical Engineers, First Edition. Jack Hipple.
© 2017 American Institute of Chemical Engineers, Inc. Published 2017 by John Wiley & Sons, Inc.

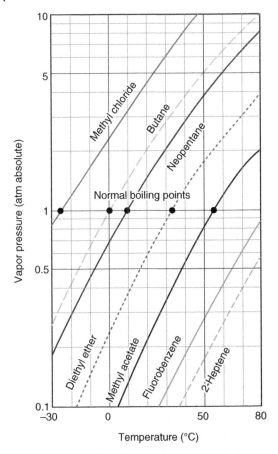

Figure 10.1 Boiling points of organic compounds. Source: Cmglee, https://commons. wikimedia.org/wiki/File:Vapor_pressure_chart.svg. Used under CC BY-SA 3.0, https:// creativecommons.org/licenses/by-sa/3.0/deed.en. © Wikipedia.

expect that if we took a mixture of ethanol and water and heated it, the over-head vapors would contain more ethanol than water. That's exactly what happens. If we were to then condense this vapor mixture, we would now have a liquid mixture with a higher concentration of ethanol than we started with. If we then boiled this mixture again, the overhead vapor would be still richer in ethanol. Condensing and reboiling would continue this progression toward a nearly pure ethanol overhead stream. Conversely, the remaining liquid would become enriched (i.e., a higher concentration) with water, eventually produc-ing a nearly pure water stream. This is the basic concept of distillation—boil, condense; boil, condense; boil, condense, etc. Done a sufficient number of times, we should be able to produce a nearly pure overhead stream of the lower

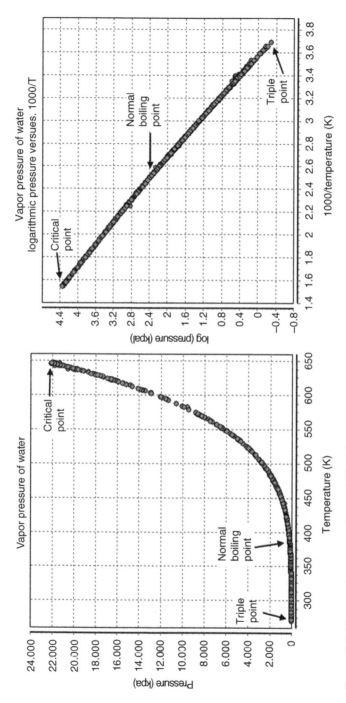

Figure 10.2 Water vapor pressure. Source: Wilfried, https://commons.wikimedia.org/wiki/File:Vapor_Pressure_of_Water.png. Used under CC BY 3.0, https://creativecommons.org/licenses/by/3.0/deed.en. ©Wikipedia.

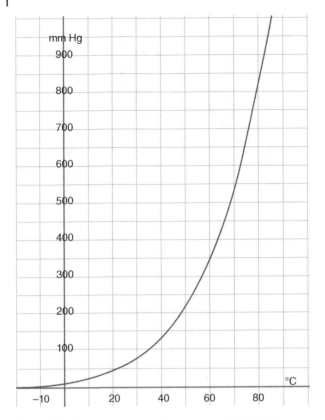

Figure 10.3 Vapor pressure of ethanol versus temperature.

boiling component and a nearly pure bottom stream of the higher boiling component as a "residue." In concept, two liquids with any difference in boiling
point can be separated by distillation, but as we will see, it is frequently not that
simple, although the basic concept is the same.

We can display this process as follows (the energy input is not shown, only
the concentration changes), shown in Figure 10.4.

Now in order for this concept to be practical, more than one "stage" is needed,
requiring the need to condense the more concentrated overhead vapor we have
produced, condense it again (liquefy it), reboil it, and so on as many times
as necessary to get the purity of the more volatile that we desire. We do this in
a *distillation column*, which incorporates the vaporization and condensation
in one process unit as shown in Figure 10.5.

Let's now discuss some of the fundamentals of distillation, the effect of
physical properties, and some of the fundamental design parameters that are
involved.

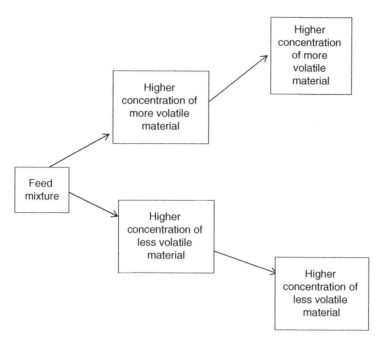

Figure 10.4 Concentrating by boiling and condensing.

Figure 10.5 Typical distillation system. Source: Sponk https://commons. wikimedia.org/wiki/File:Continuous_ Binary_Fractional_Distillation.PNG. Used under CC BY-SA 3.0, https:// creativecommons.org/licenses/ by-sa/3.0/deed.en. © Wikipedia.

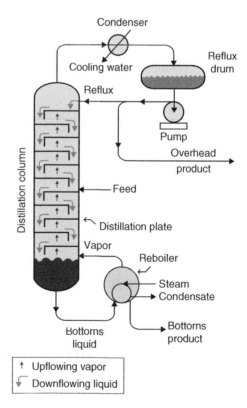

Raoult's Law

Let's assume that we have one or more liquids in solution. How do we decide how much volatility (vapor pressure) is above the liquid for any of the components? For an ideal solution, we use Raoult's law:

$$P_a = x_a P^0$$

where P_a is the partial pressure or mole fraction in the vapor space above the solution, P^0 is the vapor pressure of "a" by itself (at the temperature of concern), and x_a is the mole fraction of component "a" in the solution. This equation applies only to ideal solutions (i.e., hydrocarbon mixtures) and not to solutions with significant molecular interaction (alcohols, aldehydes, ketones, etc.). In many cases, the data must be gathered with laboratory work or found in the literature.

Regardless of whether the solution is ideal or not, the sum of the component partial pressures $(P_a + P_b + P_c + ...)$, when boiling, must equal the total pressure in the system. If the sum does not equal, then there is inaccuracy with the data or measurements of concentration, temperature, or pressure.

With insoluble liquids such as CCl_4 and water, the principle still applies; the sum of the partial pressures must equal the total pressure in the system.

We can visualize Raoult's law on this graph, for a three-component system (this equation is valid regardless of whether the solution is ideal or not) as shown in Figure 10.6.

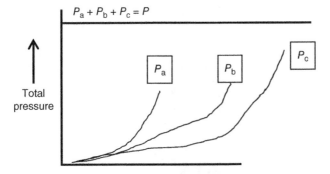

Figure 10.6 Raoult's law: total pressure = sum of partial pressures.

Another way of expressing this is that when the sum of all the partial pressures of the individual components in the mixture or solution equals the total pressure, the system will boil. The total pressure of the system will have a great effect on the boiling temperature.

Some other terms that we use to describe a volatile liquid system are the following:

Volatility is the ratio of the partial pressure of the component in the mixture divided by its mole fraction in the liquid. We can express this as

$$V_a = \frac{P_a}{X_a}$$

where V_a is the volatility of compound "a."

Relative volatility, "α," is the ratio of one component's volatility to another and is one of the key measures of how difficult it is to separate a mixture by distillation. This number tells us how difficult a distillation separation might be. If α is low, there is not much difference in the volatility of the compounds, and separation by distillation will be costly, but not necessarily impossible. If α is high, then distillation will be an economical choice, involving fewer trays (less tall column) and less energy.

We can visualize the volatility of a given compound on a simple y versus x graph, where y is the mole fraction of the component in the vapor phase versus the mole fraction in the liquid phase as shown in Figure 10.7.

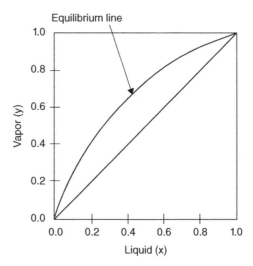

Figure 10.7 Vapor–liquid equilibrium line.

The 45° line in this graph is simply a reference line showing where y and x have the same values. The larger the distance between the equilibrium line and this reference line, the greater the volatility (y/x) of the compound.

If we go back to the original description of distillation (boil, condense; boil, condense; boil, condense), we can visualize some equipment approaches to using the difference in volatility to increase the purity of the more volatile component while at the same time increasing the concentration of the less volatile component.

Batch Distillation

This would describe a batch process where a charge of a liquid mixture is charged into a vessel with a means for supplying heat (coils or jacket). As heat increases the temperature of the mixture, the more volatile component will come over in the vapor at a much greater concentration, proportional to the α of the mixture. As the more volatile component distills over, its concentration in the charge vessel will decrease and more of the less volatile component will distill over, reducing the purity of the overhead product. At some point, the specification for the overhead product will drop below specification, and the distillation will need to be stopped. This type of distillation is frequently used to remove a solvent from a batch-reaction process. (If a solvent is the only component coming overhead, this type of operation would be more appropriately called an evaporator, to be discussed later). The overhead vapor is condensed in a condenser/heat exchanger. An example of such a process is shown in Figure 10.8.

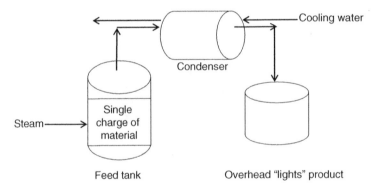

Figure 10.8 Batch distillation.

This process uses only one stage of "boil/condense" and so is suitable only for systems with a large relative volatility difference.

Flash Distillation

This is a continuous one-stage distillation process. The feed is continuous and product removals (top and bottom) are both done continuously. A simple flow sheet of this type of process is shown in Figure 10.9.

It is also likely that the feed to such a continuous one-stage flash distillation may be preheated, and it is also possible, in either of these cases, that vacuum may be used to reduce the boiling point of the solution.

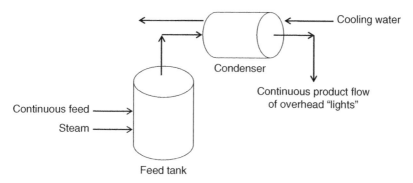

Figure 10.9 Continuous flash distillation.

Continuous Multistage Distillation

An example of a process diagram of a distillation column is shown in Figure 10.10.

As we can see, there are numerous design variables that need to be chosen. These include the method of contacting between vapor and liquid (this diagram shows what is known as a sieve tray), the internals within the column, the amount of reflux returned to the column (a reminder that unless we have reflux, we have no method for producing the "condense" mechanism to achieve more than one stage of separation), the nature and size of the heat exchangers (both top and bottom), and the height and diameter of the column.

Let's look at each of these in more detail:

1) Amount of reflux returned to the column to provide the necessary "condensing" part of the process. This will also directly affect the amount of reboiling at the bottom of the column. The choice of this ratio will affect what we call the "operating line" for the column, which is in effect, the line that graphically shows the mass balance and actual compositions of the liquid and gas flows at any particular point in the column. This is normally referred to as the *reflux ratio*, or the ratio of moles of material returned to the column divided by the moles taken off as product. We will see examples of this graphical display in the next section.
2) Method of contacting between the liquid and gas flows in the column. This can be either a tray or packings (the previous diagram shows a particular type of type). The packing material can be ceramic, plastic, or an inserted structure inside the column.
3) Nature of heat exchangers required at the top and bottom of the column. One will be a condenser (at the top, to produce product and generate reflux) and the other a reboiler (to provide the energy to boil mixtures prior to condensing).

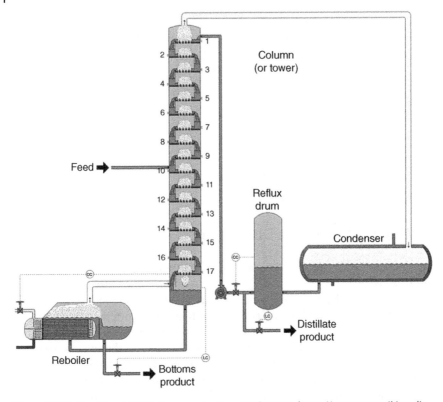

Figure 10.10 Traditional distillation process. Source: Guzman, https://commons.wikimedia. org/wiki/File:Distillation_Column_(Tower).png. Used under CC BY-SA 3.0, https:// creativecommons.org/licenses/by-sa/3.0/deed.en. © Wikipedia.

4) Diameter of the column. This will be affected by the liquid and vapor flow rates, densities and viscosities of these streams, and the ratio of liquid to gas flow rates.
5) Height of the column. This will be primarily determined by the relative volatility of the materials being distilled, the purity of the overhead or bottoms stream required, and the reflux ratio.
6) Efficiency of contacting between the liquid and vapor flows in the column. This is affected by numerous variables including the design of the contacting system and the physical properties of the liquid and vapor, including density, surface tension, and viscosity.

Reflux Ratio and Operating Line

Recall why we need reflux at all. If we did not do this and took the overhead vapor off and then condensed it as the product, this would be the equivalent of a one-stage distillation. The amount we return to the column directly affects the number of trays or height of packing required. The more we return to the column, the more "boil/condense" we do and the shorter the column needs to be. At the same time, the higher flows used require the column to be wider in diameter, and/or a larger pressure drop is seen (high liquid and gas flows = higher pressure drop). The less we return, the less "boil/condense" we get (per stage of contact), thus requiring a taller column. There is a chemical engineering optimization between the capital cost of the column, its height and diameter, the cost of the heat exchangers, and the water and steam usage. There is no one correct design for a column; it will depend on the trade-off of cost and capital, energy and cooling water demands, required purity of the top and bottoms product, and occasionally height limitations within a building.

Though most all distillation column design is done via software programs, it is useful to envision what is going on inside the column via graphical analysis, which is how columns were designed prior to software program availability.

The operating line is a graphical display of the mass balance (the liquid and vapor compositions) versus height in the column. Its slope will change based on the reflux ratio within the column as well as whether we are describing the top (rectifying) section of the column (above the feed input location in Figure 10.10) or the bottom (stripping) section of the column (below the feed input location in Figure 10.10), as shown in Figure 10.11.

The upper operating line is basically a material balance line above the feed point and into the column and the product taken off the top. The lower operating line is the material balance between the feed and bottom product from the column. The slopes of these lines and where they intersect are determined by the amount of reflux, the reboiled vapor rate at the bottom, and the temperature of the feed into the column. The junction of the two lines is where the feed is introduced into the column. If the column is operated under total reflux (all the product is returned to the column), the two lines move downward to the 45° reference line until they intersect. The distance between the operating lines and the vapor–liquid equilibrium line gives us a feel for the number of

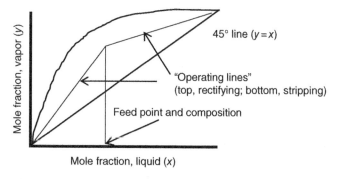

Figure 10.11 Graphical representation of distillation.

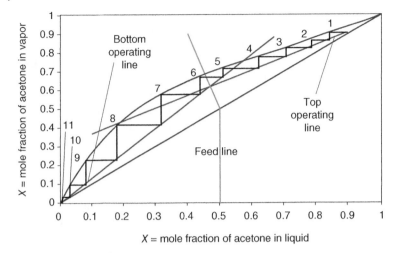

Figure 10.12 McCabe–Thiele diagram for acetone–ethanol distillation. Source: Chemical Engineering Progress, 3/12, pp. 35–41 and 6/13, pp. 27–35. Reproduced with permission of American Institute of Chemical Engineers.

stages (boil/condense) we need to produce the desired product purity at the top and the desired composition at the bottom of the column.

Such a column diagram for the separation of acetone and ethanol is shown in Figure 10.12. This type of diagram is known as a McCabe–Thiele diagram and is the best way to visualize what is occurring inside a distillation column.

At the other end of the spectrum, we could greatly increase the reflux, minimizing the number of trays and decreasing the height. However, the energy costs will go up and the column will need to be larger in diameter. The diagram for this choice is shown in Figure 10.13.

We have a trade-off between operating cost and capital to which there is no one answer. The design will be a choice based on the cost factors as viewed by the user of the column.

At the end of this extreme would be total reflux of the overhead product back to the column with no product taken off. This is frequently a start-up test done on a column during start-up and tells us the best performance the column will achieve at the same pressure. This condition maximizes the distance between the vapor–liquid equilibrium line and the "operating line," which shows the actual compositions that should exist in the column at any stage or tray. At total reflux, the operating line overlaps exactly with the 45° reference line we discussed earlier.

This optimization of reflux ratio and height of column (number of contacting devices or amount of packing) can be represented as seen in Figure 10.14.

The extremes at each end of the curve represent the minimum reflux ratio that can be used to accomplish the separation (at maximum number of trays) and the minimum number of trays required. The shape of this curve will be a strong function of the relative volatility (α) discussed at the beginning of this unit.

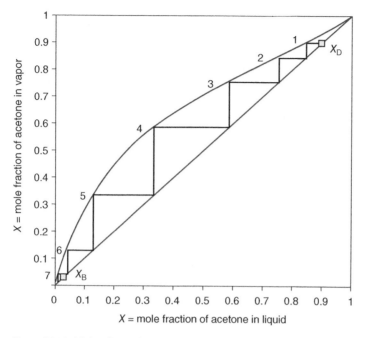

Figure 10.13 High reflux in distillation. Source: Chemical Engineering Progress, 3/12, pp. 35–41 and 6/13, pp. 27–35. Reproduced with permission of American Institute of Chemical Engineers.

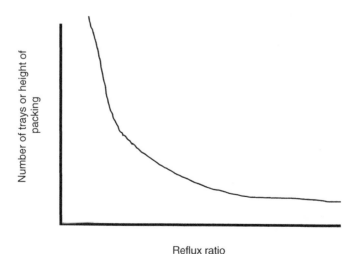

Figure 10.14 Number of trays versus reflux ratio in distillation.

Pinch Point

There is an interesting limitation at the low reflux extreme of column design. If the reflux rate is low enough, the operating/material balance line will intersect the vapor–liquid equilibrium curve as shown in Figure 10.15. Under these conditions, the column will produce no separation or change in composition as there is no driving force of composition difference. The reflux ratio, L/V ratio, vapor–liquid equilibrium data, and number of trays (or height of packing) are all connected and interact with each other in a design sense to produce a distillation column design.

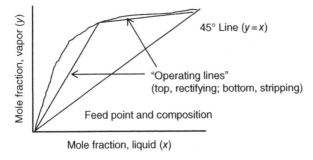

Figure 10.15 Graphical representation of a distillation pinch point.

The intersection of the lower and upper column operating lines at the vapor equilibrium line eliminates the driving force for separation. This graph represents the opposite of total reflux, where we don't reflux enough material back into the column to provide a driving force between the operating line and the vapor–liquid equilibrium line. The column becomes inoperable (but uses minimal energy and water to condense) in the same sense that a column on total reflux produces no product but uses the shortest tower.

Feed Plate Location

It is always best to introduce the feed into the column at a point where we calculate that composition to exist when the column is operating. This minimizes the energy and capital cost of "re-equilibrating" the feed to the conditions in the tower.

Column Internals and Efficiency

In all the discussions up until now, we have assumed that if the graphical design shows a certain number of trays, then that's the number we install. The real world is different. We are basically contacting liquids and gases in counter current flow using some kind of physical contacting mechanism. The efficiency of this contacting is affected by a large number of process and property variables:

1) Physical Properties. It is easy to visualize how differences in density and viscosity of the gases and liquids could affect how efficiently and completely a liquid and a gas would mix on a plate or within the voids of a packed bed. These properties may vary significantly from top to bottom of the column due to temperature gradients from top to bottom (recall the effect of temperature on viscosity discussed earlier). The density difference between the gas and the liquid, including those from top to bottom, will also affect the speed of mixing and equilibration. The graphs shown previously assume that perfect equilibrium is achieved at each stage prior to the next stage of contacting.

2) Column Internals. We can specify a certain number of trays or height of packing on the assumption that we get excellent and complete contacting between gas and liquid, but there are a number of reasons why this does not happen. These include plugging of holes in trays or plates with process contaminants and flow restrictions in any hydraulic or gas pathways. Also of concern is slow degradation in any support equipment holding trays, and packing in place.

3) Column Efficiency Measurements. The measurement of column efficiency can be done in a variety of ways. The first would be overall plate efficiency. We measure the actual performance in a pilot plant, and compare it to the predicted performance from previous calculations. For example, if the graphs or calculations show that 7 trays should be needed and 10 are actually needed, we can say that the overall tray efficiency is 70% and use this as a design factor for similar systems using similar materials. Another is to try to determine the actual efficiency of each individual plate or height of packing. This is much more appropriate if we expect large differences in density, viscosity, and liquid/gas ratio over the entire height of the column. This efficiency, known as the Murphree plate efficiency, would be integrated over the entire column height to determine the actual number of required trays.

We have mentioned occasionally the use of loose packing instead of trays. We will discuss this further, but another type of efficiency measurement is called the height equivalent to a theoretical plate (HETP). This is in effect a statement that x number of feet of packing provides the same degree of contact as one physically distinct tray.

Unique Forms of Distillation

There are several unique types of distillation used for various reasons:

1) Steam Distillation. Occasionally, an overhead product is desired that decomposes close to or below its normal boiling point that would exist at the top of a distillation column. If the product is not water soluble, steam can be introduced as a diluent, artificially lowering its boiling point.

2) Vacuum Distillations. Operating a distillation under vacuum will lower the boiling points and vapor pressures within the column. A vacuum distillation can produce the same effect as a steam distillation. With steam, there is an energy cost, increased capital costs, and a need to separate the product from the overhead steam. The vapor piping systems are also larger and more costly. Vacuum systems also cost more than atmospheric or pressure distillations.

3) Azeotropes and Azeotropic Distillation. Most of the time, azeotropes are considered barriers to achieving a high purity material, but if an azeotrope is identified that can withdraw an undesired compound (this is what is done by adding benzene to the 95% ethanol/water azeotrope) to a sufficient degree, then this can be a way around a limitation to achieving a desired purity of the product. A typical y versus x or vapor–liquid equilibrium diagram for two liquids has a diagram similar to what we have already seen (as in Figure 10.16).

An *azeotrope* is a composition point where the boiled vapor has the same concentration as the liquid. This can occur when there are strong molecular interactions between liquid components. Combinations of polar compounds (water, alcohols, ketones, aldehydes) are most likely to exhibit this behavior. For normal hydrocarbons

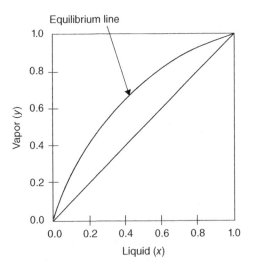

Figure 10.16 Vapor–liquid equilibrium for an ideal solution.

this is unlikely. It is best, however, not to assume anything and look up azeotropic compositions in available literature references. When an azeotrope occurs, it is not possible to obtain further change in composition, beyond this point, by boiling and condensing since the liquid and vapor have the same composition. In other words the *y* and *x* on the mole fraction diagrams we have displayed so far have the same value.

Azeotropes are, in effect, significant deviation from Raoult's law. The deviations can be either "positive" (the total component vapor pressure is more than calculated from Raoult's law) or "negative." If Raoult's law was represented as a straight line, then azeotropes would be represented as seen in Figure 10.17.

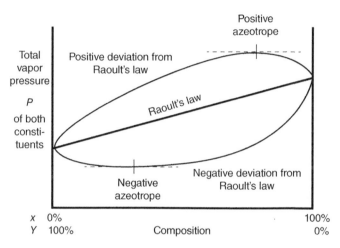

Figure 10.17 Azeotropic compositions: positive and negative. Source: Karlhahn, https://commons.wikimedia.org/w/index.php?curid=2685348. © Wikipedia.

An illustration of such a system, propanol–water, is shown in Figure 10.18. The azeotropic composition is around 70 mole%.

A similar azeotrope is seen for ethanol and water at 95.5% ethanol (Figure 10.19). A maximum boiling azeotrope will have a boiling point higher than predicted from Raoult's law and a minimum boiling azeotrope will have a boiling part lower than predicted. We can see this as we plot the boiling point of azeotropic mixtures versus composition. We will see either a positive deviation (higher BP) or negative deviation (lower BP) from an ideal solution calculation as shown in Figures 10.20 and 10.21.

Very frequently we will find that commercial products, such as ethanol and nitric acid, are sold and shipped at their azeotropic composition points (95.5% and 68%, respectively), since further changing the concentrations will involve added cost and complexity.

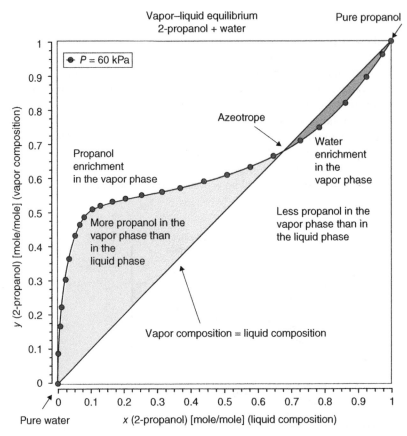

Figure 10.18 Azeotropic composition for a nonideal system. Source: Wilfried, https://commons.wikimedia.org/w/index.php?curid=9788841. Used under CC BY-SA 3.0, https://creativecommons.org/licenses/by-sa/3.0/deed.en. © Wikipedia.

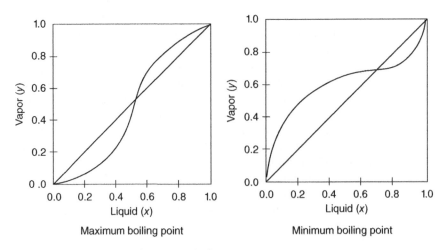

Figure 10.19 Maximum and minimum boiling azeotropes.

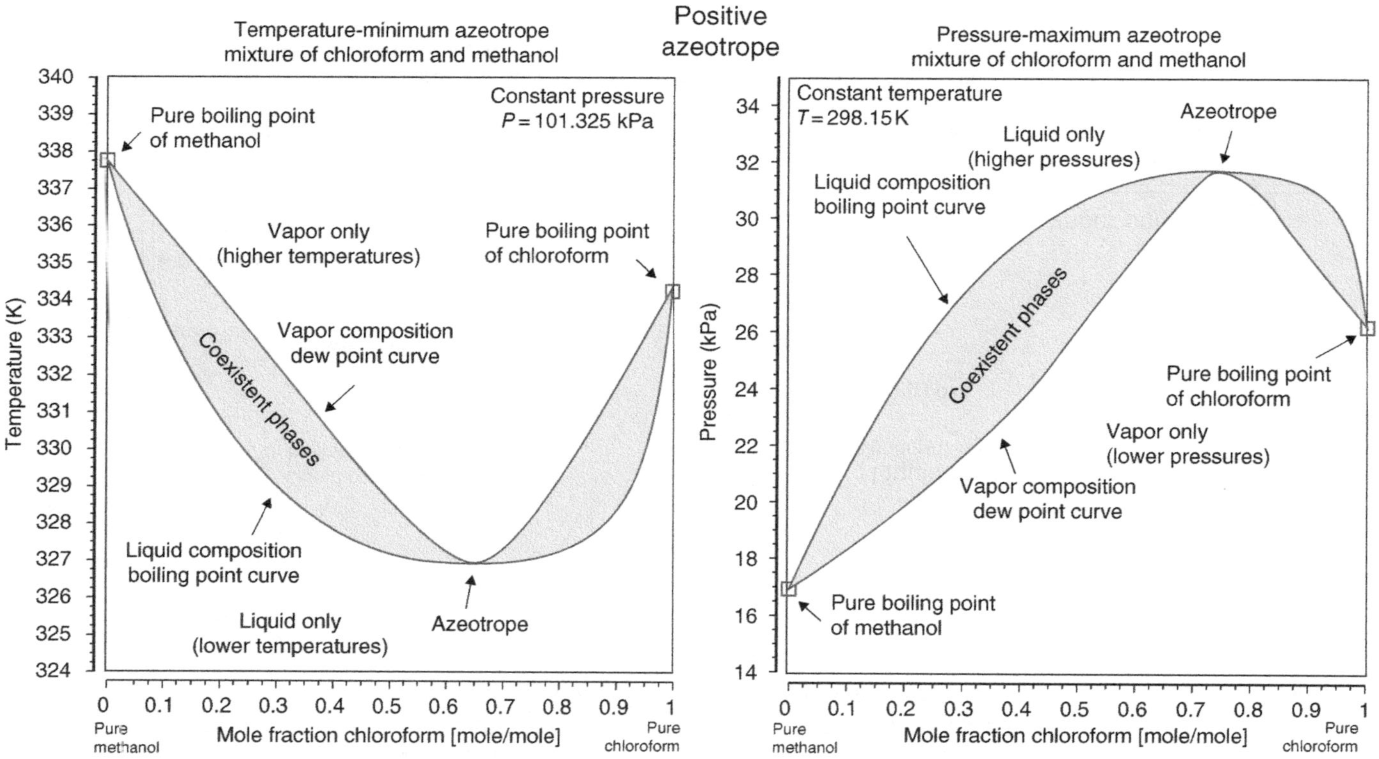
Positive azeotrope
Temperature-minimum azeotrope mixture of chloroform and methanol
Constant pressure P = 101.325 kPa
Pure boiling point of methanol
Vapor only (higher temperatures)
Pure boiling point of chloroform
Vapor composition dew point curve
Coexistent phases
Liquid composition boiling point curve
Liquid only (lower temperatures)
Azeotrope
Temperature (K)
340
339
338
337
336
335
334
333
332
331
330
329
328
327
326
325
324
0 0.1 0.2 0.3 0.4 0.5 0.6 0.7 0.8 0.9 1
Pure methanol
Mole fraction chloroform [mole/mole]
Pure chloroform
Pressure-maximum azeotrope mixture of chloroform and methanol
Constant temperature T = 298.15 K
Azeotrope
Liquid only (higher pressures)
Liquid composition boiling point curve
Coexistent phases
Pure boiling point of chloroform
Vapor only (lower pressures)
Vapor composition dew point curve
Pure boiling point of methanol
Pressure (kPa)
34
32
30
28
26
24
22
20
18
16
14
0 0.1 0.2 0.3 0.4 0.5 0.6 0.7 0.8 0.9 1
Pure methanol
Mole fraction chloroform [mole/mole]
Pure chloroform

Figure 10.20 Positive azeotropes *T*–*x*–*y* diagram. Source: Wilfried, https://commons.wikimedia.org/w/index.php?curid=7437968. Used under CC BY-SA 3.0 https://creativecommons.org/licenses/by-sa/3.0/deed.en. © Wikipedia.

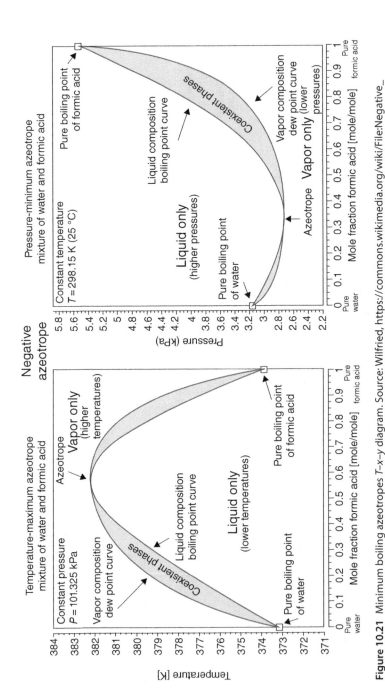

Figure 10.21 Minimum boiling azeotropes *T*–*x*–*y* diagram. Source: Wilfried, https://commons.wikimedia.org/wiki/File:Negative_Azeotrope.png. Used under CC BY-SA 3.0, https://creativecommons.org/licenses/by-sa/3.0/deed.en. © Wikipedia.

A way of dealing with an azeotropic system, if it is desired to produce a product beyond this composition, when the lighter boiling component is the limiting factor is to change the pressure in what is known as "pressure swing" distillation. In this situation, we change the pressure (moving toward either vacuum or pressure) to change the boiling point of the mixture, allowing the azeotropic point to be changed.

An example of such a process is shown in Figure 10.22.

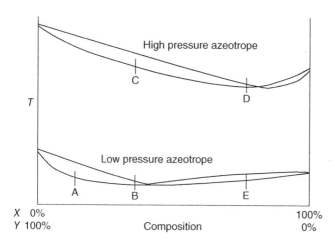

Figure 10.22 Using pressure change to break an azeotrope.

The separation of the 95–5% water azeotrope is a commercial example of the use of this technique, varying the pressure between atmosphere and 8 atmospheres to "jump" over the normal atmospheric azeotrope at 95%. Another approach to the azeotrope challenge, when one of the components is water, is to add salt to the water, raising its boiling point.

Another interesting way of using azeotropes is in a positive way, that is, introduce a third component that forms a multiple component azeotrope to remove a component from a binary mixture that is difficult to separate. Acetic acid and water have very close boiling points, and the introduction of ethanol to deliberately form the water/ethanol azeotrope discussed earlier removes the water and makes it easier to produce pure acetic acid.

Multiple Desired Products

In many cases, there is a feed mixture that needs to be separated into multiple streams of pure products. There are two basic approaches to this. One is, with sufficient knowledge of the vapor pressures as a function of composition, to withdraw side products at various heights within the column. This concept is used in separating the basic components of crude oil as illustrated in Figure 10.23. This illustration shows not only the distillation but also the uses for the various products. As we would expect, the lower boiling components are in the upper section of the column (LPGs, hydrocarbon feedstocks, gasoline) and the higher boiling components in the bottom half (diesel fuel, lubricants, asphalt factions, etc.).

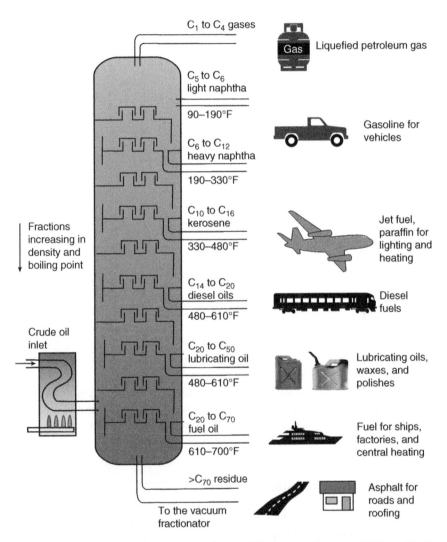

Figure 10.23 Crude oil distillation. Source: Chemical Engineering Progress, 11/12, pp. 32–38. Reproduced with permission of American Institute of Chemical Engineers.

Though not as well known by the public, the separation of air into its components such as oxygen, nitrogen, argon, carbon dioxide, and rare gases such as krypton is done in the same way, although the operating temperatures in the columns are several hundred degrees below zero and large amounts of insulation are used around process equipment.

Column Internals and Efficiencies

Regardless of what kind of distillation we are running, decisions about the type and method of contacting need to be made. As mentioned several times already, there is rarely one choice but a decision based on cost and practical aspects relating to the fluids and vapors being distilled, separated, and collected. The contacting can be done via trays or loose fill packing.

Tray Contacting Systems

There are basically three types of contacting trays: bubble cap, valve, and sieve:

1) Bubble cap trays are probably the oldest commercially used trays and allow the vapor rising up the column to "bubble through" the downflowing liquid to create the contacting, as shown in Figure 10.24.

    This type of tray provides intimate mixing but can be costly to fabricate. With dirty or high viscosity fluids, the ability of the cap to float up and down can be hindered, and cleaning can be a major challenge. If a cap is plugged off in some way, the vapor pressure drop will increase across the tray as we have the same amount of vapor trying to rise through fewer holes. Raising the pressure or pressure drop will affect the vapor–liquid equilibrium for the column as well as peripheral equipment performance. One excellent performance characteristic of these types of trays is their ability to minimize "weeping." This refers to the liquid on the tray to "weep" down into the tray below. Anything that promotes mixing between the trays defeats the whole purpose of distillation, which is to separate the vapor and liquid in distinct stages

2) Another type of tray is a sieve tray, which is basically a plate with holes in it, designed to have both the downcoming liquid and rising vapor go through the same hole, creating intimate mixing. This type of tray, shown in Figure 10.25, due to its low cost and ease of maintenance, has become the workhorse of the chemical industry.

    The major operating issue with sieve trays is the potential for weeping. Since there is no mechanical barrier to liquid downflow, as in the bubble cap design, there is greater possibility of plate-to-plate mixing, lowering the

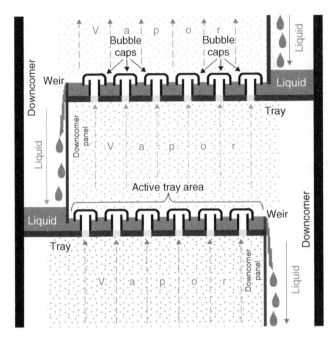

Figure 10.24 Bubble cap trays. Source: Pedlackes, https://commons.wikimedia.org/wiki/File:Bubble_Cap_Trays.PNG. Used under CC BY-SA 2.5, https://creativecommons.org/licenses/by-sa/2.5/deed.en. © Wikipedia.

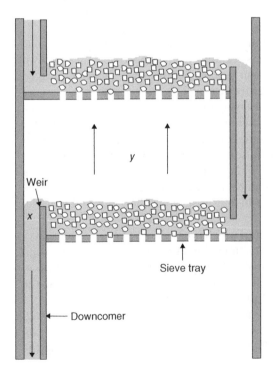

Figure 10.25 Sieve tray. Source: Reproduced with permission of Brigham Young University.

efficiency of the column's separation capability. Weeping is a much greater possibility at low operating rates as there may be insufficient upward vapor flow to maintain a liquid level on the tray.

Design variables for this type of tray include the spacing between trays, the height of the weir, the hole size, and the geometry of the downcomer. These will be affected by liquid and gas properties (density, viscosity) and the surface tension and tendency to foam. To repeat, any operating factor that causes mixing between trays defeats the entire purpose of the process, which is to separate the boiling and condensing aspects.

Another type of tray design is the valve tray, a hybrid between the bubble cap and sieve tray. Here the sieve tray holes have a mechanical sealing device internal to the tray, performing somewhat the same function as a bubble cap but in a less costly way to manufacture (see Figure 10.26).

Figure 10.26 Valve tray. Source: Chemical Engineering Progress, 11/12, pp. 32–38. Reproduced with permission of American Institute of Chemical Engineers.

The "weeping" that would allow liquid to drop down through a tray opening and mix with the tray below can be influenced by a variety of fluid and process variables including vapor velocity, liquid density, liquid depth on the tray, and the differences in liquid and gas properties such as density and viscosity.

Another type of tray is a dual flow tray, which is usually angularly designed to allow gas and liquid to go through the same holes.

Packed Towers in Distillation

Instead of a discrete separation of gas/liquid contacting, it is also possible to have the gas and liquid flowing up and down the tower in a continuous way by filling the tower with loose fill packing. Examples of such packings are both random and structured. Random packings include Raschig rings, Pall rings, Intalox™ saddles, and a wide variety of other preformed ceramic, metal, and plastic shapes.

Figure 10.27 Tower packings. Source: Used with permission of Sulzer AG.

Packed towers used for distillation can have several advantages:

1) Lower Pressure Drop. This can mean easier separation of close boiling components and minimize the need for vacuum to separate heat-sensitive materials. This can also minimize the cost of vacuum equipment.
2) Materials from which loose fill packing can be manufactured, in combination with a corrosion resistant lined column, can better handle corrosive chemical systems.

There are a few disadvantages as well. With the large surface area, there is a tendency for liquid coming down the column to tend to gravitate toward the walls, so it is usually necessary to provide redistribution every 10–20 ft. down the column. The liquid is collected from the walls and redistributed to the center of the column. The degree to which this happens will be affected by density, viscosity, and surface tension of the liquid. Second, if ceramic packing is used, it is critical that, during start-up, the boxes in which this type of packing is normally shipped be gently dropped via a rope and gently tipped over. Ceramics are very brittle materials and, if dropped from a significant height, will shatter upon reaching the bottom of the column, resulting in small pieces that can clog support trays at the bottom of the column and cause excessive pressure drop.

Structured packing is a way of inserting fixed geometric shapes within the diameter of a column, as shown in Figure 10.28. These types of packing inserts have extremely low pressure drop and are highly efficient but also are more expensive than either tray or loose fill packed towers for distillation. Given their high efficiency and low pressure drop, they can provide

Figure 10.28 Structured packing. Source: Used with permission of Sulzer AG.

separation capability in situations where very low relative volatilities exist, height limitations on columns may exist, or product decomposition is of concern.

We have mentioned pressure drop across tower packings a number of times. Pressure drop across distillation towers, from large to small, would be:

1) Bubble cap
2) Valve and dual flow
3) Sieve trays
4) Loose packing
5) Structured packing

There are some practical issues in the operation and design of packed towers, both for distillation as well as absorption and stripping (to be discussed in Chapter 11):

Plumb tests must be done during start-up. This is more important with a packed tower than a tray tower. With the much higher surface area, there is a tendency for the downflowing liquid to gravitate toward the walls vs. the direction it first starts. This is also why occasional redistribution of liquid flow is required in a packed tower.

Leveling tests should also be done on tray towers for the same reason.

Support plates for packed towers must be unobstructed to avoid high pressure drop. With ceramic packings, due to their inherent brittleness, installation must be done carefully by gently tipping over the typical several cubic feet boxes at the bottom (or their next layer) and dump them from the top of the tower.

Distillation reboilers We have already reviewed the basics of heat transfer and heat exchangers. At the bottom of a distillation tower, heat input is required to boil up the heavy liquid and provide the "boil" part of the boil/condense distillation mechanism. A general diagram of this type of exchanger is seen in Figure 10.29.

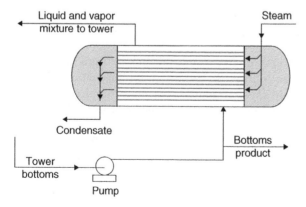

Figure 10.29 Reboiler configuration. Source: Mbeychok, https://commons.wikimedia.org/wiki/File:ForcedCirculation.png. Used under CC BY-SA 3.0, https://creativecommons.org/licenses/by-sa/3.0/deed.en. © Wikipedia.

It may also be possible, with a low heat requirement, to supply the reboiling needed with a jacket around the bottom of the distillation column or by tubes inserted into the bottom of the column. The design of these types of heat exchangers is fundamentally no different than we discussed earlier.

Summary

Distillation is a unique unit operation that allows us to separate liquids based on their differences in vapor pressure and boiling points. There is a wide range of design choices, both in the types of internals and contacting devices, as well as an optimization of the physical and geometric parameters of the column. Exact choices will be affected by capital and energy cost variables, geometric limitations, and the current and future quality specification requirements for the various products we can produce via this chemical engineering unit operation.

Coffee Brewing and Distillation

The way we normally brew coffee does not involve distillation processes in any way, so we'll come back to this example in the next chapter.

Discussion Questions

1   How many distillation processes are run in your complex? What is their purpose? How many are run with varying compositions of feed material? How is the operation varied in response to these changes?

2   How is the reflux ratio determined and controlled? How do limitations in cooling water affect the operation of overhead condensers?

3   What are the specifications on the overhead and bottoms products? How tightly must they be controlled? How are the reflux ratio and feed rates varied to stay within specifications? What opportunities might be presented if a higher purity overhead product could be produced?

4   How was the current mechanical design (tray type, packing, condenser, and reboiler types) determined? Have these decisions been reevaluated over time?

5   Is the minimum reflux ratio known?

6   What are the impacts of changing utility pressures and temperatures?

Review Questions (Answers in Appendix with Explanations)

1 Distillation is a unit operation based on differences in:
  A __Solubility
  B __Density
  C __Vapor pressure
  D __Crystallinity

2 A material, solution, or mixture boils when:
  A __The solution is rolling around and bubbling violently
  B __The sum of all the partial pressures equals the total pressure
  C __It's mad
  D __The partial pressures exceed the external pressure by 10%

3 The key determinant in how easy it is to separate a mixture by distillation is the:
  A __Volatility of the relatives of the mixture
  B __Whether the relatives want to be separated
  C __Ability to heat selectively the most volatile component
  D __Relative volatility of its various components

4 On a graphical plot of a distillation system, the 45° line represents:
  A __The vapor phase and liquid phase having the same composition
  B __All phases are created equal
  C __One component has 45% more volatility than another
  D __One component has 45% less volatility than another

5 In a two-component distillation system where the relative volatility is displayed on a y–x graph, a higher relative volatility will be displayed, versus a 45° line, as:
  A __No difference in the lines
  B __A small difference in the lines
  C __A large difference between the lines
  D __Price of company, suppliers, and customer stocks that change minute by minute

6 In a batch distillation system, the maximum number of stages of separation possible is:
  A __One
  B __Depends on relative volatility
  C __A function of heating rate
  D __A function of the batch size

7 In a conventional continuous distillation system, the top of the column will always contain:
A __A higher concentration of the less volatile component
B __A higher concentration of the more volatile component
C __A higher concentration of the less dense material
D __A higher concentration of the material desired by the customer

8 Returning reflux to a distillation column allows:
A __More energy to be wasted
B __More cooling water to be wasted
C __Multiple vaporizations and condensations, yielding purer top and bottom products
D __More capital expenditures to be wasted on a reboiler and condenser

9 Increasing reflux to a distillation column results in:
A __Higher pressure drop
B __Purer overhead product
C __More cooling water to be used
D __All of the above

10 Decreasing reflux to a distillation column results in all except:
A __Less cooling water and reboiler steam use
B __Lower overhead purity
C __Less intensive process control
D __Pressure drop across the column decreases

11 The "operating line" of a distillation column represents a graphical display of:
A __The line drawn by the process operators when the process computers are offline
B __The mass balance within the column
C __The line of code that operates the column
D __The line that no one on the operating floor is allowed to cross

12 Varying the reflux ratio in a distillation column allows us to:
A __Adjust the quality of overhead and bottom products
B __React to changes in feed compositions
C __Allow process adjustments to upstream and downstream processes
D __All of the above

13 Vacuum distillation can result in all but:
A __Increased energy use
B __Separation of azeotropes
C __Separation of high boiling components
D __Smaller distillation columns

14 Azeotropes are:
   A __Special mixtures of chemicals that come from the tropics
   B __Mixtures of chemicals with close boiling points
   C __Mixtures of materials whose vapor composition when boiled is the same as the starting liquid
   D __Impossible to separate

15 Ways of separating azeotropes include:
   A __Changing pressure
   B __Using an alternative separation technique
   C __Adding a third component that shifts the vapor–liquid equilibrium
   D __All of the above

16 Bubble cap trays in distillation columns have this key advantage:
   A __They trap bubbles
   B __Prevent liquid from dropping down on to a lower tray without contacting vapor
   C __Relatively expensive
   D __High pressure drop

17 Sieve trays have this disadvantage:
   A __Low pressure drop
   B __Inexpensive and easy to fabricate
   C __Can allow weeping and mixing between stages
   D __Can allow low molecular weight materials to leak through

18 Loose packings used in place of trays:
   A __Will usually have lower pressure drop
   B __Can be more corrosion resistant
   C __Are more likely to breakup due to mechanical shock
   D __All of the above

Additional Resources

Bouck, D. (2014) "10 Distillation Revamp Pitfalls to Avoid" *Chemical Engineering Progress*, 2, pp. 31–38.
Gentilcore, M. (2002) "Reduce Solvent Usage in Batch Distillation" *Chemical Engineering Progress*, 2, pp. 56–59.
Hagan, M. and Kruglov, V. (2010) "Understanding Heat Flux Limitations in Reboiler Design" *Chemical Engineering Progress*, 11, pp. 24–31.
Kister, H. (2004) "Component Trapping in Distillation Towers: Causes, Symptoms and Cures" *Chemical Engineering Progress*, 8, pp. 22–33.

Kister, H. *Distillation Design* (New York: McGraw-Hill, 1992).

Perkins, E. and Schad, R. (2002) "Get More Out of Single Stage Distillation" *Chemical Engineering Progress*, 2, pp. 48–52.

Phimister, J. and Seider, W. (2001) "Bridge the Gap with Semi-Continuous Distillation" *Chemical Engineering Progress*, 8, pp. 72–78.

Pilling, M. and Holden, B. (2009) "Choosing Trays and Packings for Distillation" *Chemical Engineering Progress*, 9, pp. 44–50.

Pilling, M. and Summers, D. (2012) "Be Smart about Column Design" *Chemical Engineering Progress*, 11, pp. 32–38.

Summers, D. (2010) "Designing Four-Pass Trays" *Chemical Engineering Progress*, 4, pp. 26–31.

Vivek, J.; Madhura, C. and O'Young, L. (2009) "Selecting Entrainers for Azeotropic Distillation" *Chemical Engineering Progress*, 3, pp. 47–53.

White, D. (2012) "Optimize Energy Use in Distillation" *Chemical Engineering Progress*, 3, pp. 35–41.

AIChE Equipment Standards Testing Committee (2013) "Evaluating Distillation Column Performance" *Chemical Engineering Progress*, 6, pp. 27–35.

http://www.che.utah.edu/~ring/Design%20I/Articles/distillation%20design.pdf (accessed on August 31, 2016).

http://www.hyper-tvt.ethz.ch/distillation-binary-reflux.php (accessed on August 31, 2016).

http://www.aiche.org/system/files/cep/20130627.pdf (accessed on August 31, 2016).

http://www.separationprocesses.com/Distillation/DT_Chp04n.htm (accessed on August 31, 2016).

11

Other Separation Processes

Absorption, Stripping, Adsorption, Chromatography, Membranes

These mass transfer operations are also a key part of the chemical engineering tool kit for specific applications. Absorption and stripping are opposites of each other, while extraction, membranes, chromatography, adsorption, and leaching are more specialty processes typically used for unique separation needs.

Absorption

Absorption is defined as the transfer of a gas into a liquid. This is done to recover a valuable component in a gas stream for reuse or sale or the removal of a component in a gas stream to eliminate an environmental discharge. If the purpose of the absorption is to remove an air contaminant prior to discharge of an airstream into the atmosphere, the absorption process may be referred to as "scrubbing," implying the need to remove something prior to discharge. A general diagram of such a process would be as shown in Figure 11.1.

Figure 11.1 shows the use of packing (as designated with an "X" across the diameter of the tower), as opposed to trays, but either can be used. Packing, frequently made from ceramic materials, is very commonly seen in this unit operation because many of the gases that require absorption or scrubbing tend to be acidic gases such as SO_2 and HCl. Spray towers, without internal packing, can also be used for highly soluble gases.

One of the key design parameters is the solubility of the gas in the absorption or scrubbing liquid (frequently water, but another absorbent or solvent could also be used such as an alkali or oil). This is a physical property variable known as Henry's law constant ("H"). "H" is the ratio of the gas concentration divided by the concentration of that same gas in the absorbing/scrubbing fluid. H frequently uses the units of atm/mole fraction. It is a valid equation only for

Chemical Engineering for Non-Chemical Engineers, First Edition. Jack Hipple.
© 2017 American Institute of Chemical Engineers, Inc. Published 2017 by John Wiley & Sons, Inc.

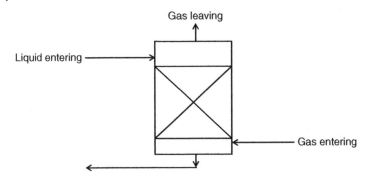

Figure 11.1 Absorption tower.

extremely dilute solutions and that is the context we will use in this chapter. In equation form, we can express the partial pressure of the gas (p) as a function of its concentration in the liquid, c, times Henry's law constant (H), that is, $p = Hc$.

At high concentrations, this relationship is not linear as it is with low concentrations.

Here are some of these values for the gases CO_2, CO, and H_2S (atm/mole fraction):

Table 11.1 Henry's law constants.

T, °C	CO_2	CO	H_2S
0	728	35 200	26 800
5	876	39 600	31 500
10	1040	44 200	36 700
15	1220	48 900	42 300
20	1420	53 600	48 300
25	1640	58 000	54 500
30	1860	62 000	60 900
35	2090	65 900	67 600
40	2330	69 600	74 500
45	2570	72 900	81 400
50	2830	76 100	88 400
60	3410	82 100	103 000

Source: Average of public information.

The following are a few points about the data in Table 11.1:

1) The Henry's law constant is very sensitive to temperature.
2) Gas solubility *decreases* dramatically with temperature. (The hotter the fluid, the less capability it has to absorb or hold the gas.) We see this every day when watching what happens to a carbonated beverage when left out in the kitchen after it has been in the refrigerator for days. It loses its "fizz," which is the carbon dioxide (CO_2) coming out of solution as the temperature rises. In chemical engineering terms, its Henry's law constant has risen by a factor of 2–4.
3) The varying slopes of these Henry's law constant values are different. H_2S is more soluble than CO at 0°C, but less soluble at 40°C. It would be a mistake to take data for two different gases, showing one more soluble than another, and assume that this difference in solubility applied at all temperatures.
4) Many gases, especially in the acid gas category such as HCl, have significant heats of absorption when they dissolve in water. It is important to calculate a heat balance on the system and estimate the temperature rise that may occur during absorption. Any increase in temperature will decrease the gas solubility.

We use a similar graphical technique to analyze a gas absorber/scrubber as we did in distillation, with the exception that there is no condensation of the overhead product (the clean gas). Instead of "plates" we frequently use the term "transfer units" to describe how tall an absorption/scrubber tower needs to be. This is a way of generalizing the number of feet of packing representing the equivalent of one contact tray. We have some of the same type of limitations we had in distillation:

1) We need to provide absorption/scrubbing fluid rates, at a minimum, above the solubility of the gas in the liquid.
2) We need to take into account the temperature rise that may occur when doing this, due to heat of absorption of the gas into the liquid.
3) There is a balance or optimization between the diameter of the column (it will increase as we increase liquid and gas velocities) and height of the column (it will decrease with higher liquid rates). As in distillation, there is no one answer but an optimization based on energy, capital, and possibly water availability and disposal costs.

Figure 11.2 shows a graphical representation of an absorber/scrubber unit with "Y" referring to the gas phase concentration and "X" referring to the liquid phase composition.

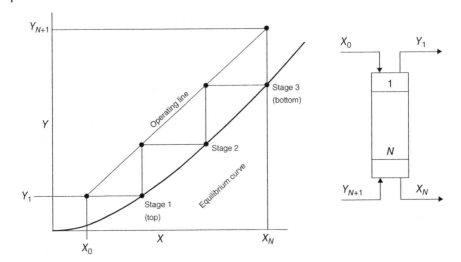

Figure 11.2 Analysis of an absorber.

The equilibrium curve shown in Figure 11.2 is basically a plot of Henry's law constant from the top to bottom of the column, indicating the amount of gas (mole fraction) that can be absorbed. The "operating line" is the same as in a distillation column, a line plotting the mass balance in the column. Since we want a component in the gas stream to move into the liquid, there must be a driving force to accomplish this. The greater the amount of absorbing/scrubbing fluid, the greater the distance between the two lines, and the number of "stages" required is fewer. In gas absorption, we generally use the phrase "transfer units" as opposed to stages; however, if we were to use a tray column to do the absorption, then we would use the same terminology. If we were to use a lower absorber fluid rate to accomplish the same degree of absorption, the tower would be taller, or we would need a more efficient packing. In the use of loose packings, vendors (or internal company experience) supply "height of a transfer unit" (HTU) equivalents, that is, 3 ft of packing type "A" is equivalent to 1 HTU. This value will vary with flow conditions and the nature of the fluid and gas properties.

Packing vendors will usually supply much of this information, but proprietary information in large corporations is also common.

We mentioned earlier the potential effect of a significant heat of solution when a gas dissolves in the absorption fluid. If we have such a situation, the equilibrium line will not be a straight line (meaning constant temperature), but will curve as its temperature increases, as shown in Figure 11.3.

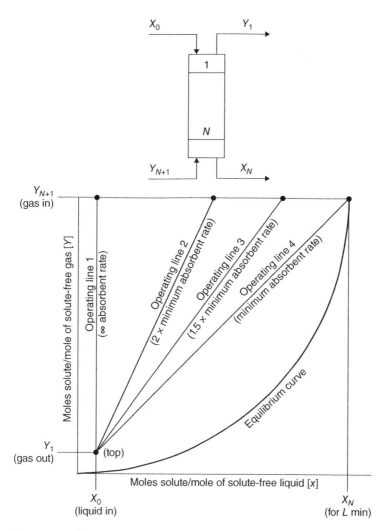

Figure 11.3 Effect of temperature on absorption.

In this type of case, a colder fluid may be needed to do the absorption required, internal cooling within the column supplied, or a taller tower required to achieve the same absorption efficiency.

Since any particular material will have a different solubility in an absorption fluid, another use of this process is to selectively remove or recover one of multiple components from a gas stream. One use of this technique is to use a hydrocarbon stream, such as a lean oil, to selectively recover low boiling hydrocarbons such as ethane, propane, butane, and pentane. If one or more of these compounds has significantly more solubility than another, it can be selectively removed or recovered.

There are occasions where the solubility of the gas may be high enough that it is not necessary to have multistage contact, and we can use what is known as a spray tower. The absorption fluid may also be a slurry of suspended solids as is often done with spray towers used in the power industry to recover sulfur dioxide (SO_2) from a power plant stack into lime ($Ca(OH)_2$) slurry.

Stripping/Desorption

This is the exact opposite of absorption or scrubbing. In this case, we have a component in the liquid (possibly material dissolved in water that cannot flow into a public waterway or company waste treatment facility) that must be removed. It is also possible that there may be a solvent in a discharge stream that has enough value to warrant recovery. In this case, the diagram for the process is exactly the same as the absorption/scrubbing process, but in these cases, the incoming liquid contains the material to be removed or recovered and the gas leaving is not clean, but now contains the material "stripped" from the liquid.

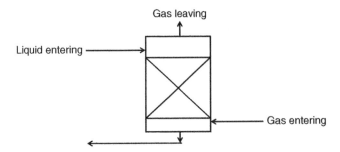

Figure 11.4 Stripping tower.

Auxiliary equipment choices are important in the design of absorbers and strippers. We have already discussed support plates and the importance, especially in packed towers, of an exact vertical installation. With stripping we have an additional concern in that we do not want fine mist of liquid being transported out the top of the column, both from an economic and environmental standpoint. We use *demisters* to accomplish this. A demister or mist eliminator is a mechanical device with very fine wire diameter that allows small liquid particles to coalesce into large enough particle size to drain back into the stripping vessel:

Figure 11.5 Typical demister. Source: Courtesy of Sulzer AG.

A widely used process in the natural gas industry illustrates both absorption and stripping. Much of the natural gas supply in the world contains not only hydrocarbons such as methane (CH_4) but also impurities such as hydrogen sulfide (H_2S) and carbon dioxide (CO_2). These two impurities not only dilute the fuel value of the natural gas but also, in the case of H_2S, cause a safety hazard due to the toxicity of this gas. However, it can be used as a feedstock for the manufacture of elemental sulfur. A class of compounds, alkanolamines, has great affinity for these two compounds. There are many different classes of these compounds (refer back to Figure 4.7 for a few examples). These compounds are used, in a combination of absorption and stripping, to produce "sweet" (non-H_2S- and CO_2-containing) natural gas that is used by both industrial and homeowners to heat their building and houses. Figure 11.6 shows an overview of this process.

In this process, sour gas is contacted with an amine that "absorbs" the sour gas impurities, H_2S and CO_2, producing a clean natural gas stream that can be utilized as a fuel. The amine, which has absorbed these impurities, frequently referred to in this industry as a "rich" amine, now needs to be "stripped" of these impurities so that it can be reused. This is done in the second tower seen in this flow sheet, producing a concentrated stream of H_2S and CO_2. This concentrated gas stream can now be used as a feedstock to a sulfur plant or a sodium hydrosulfide plant. Note that the pressures and temperatures and pressures in this process will vary greatly depending upon the temperature and pressure of the feed gas, the type of amine absorption fluid used, the level of impurities in the sour gas feed, and the intended use of the removed sour gas components.

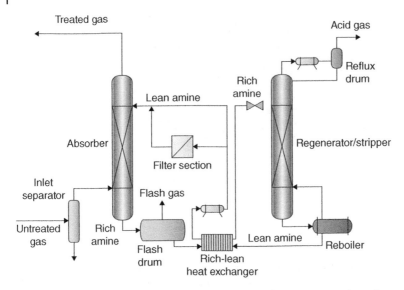

Figure 11.6 Combined absorption and stripping in purifying sour natural gas. Source: Chemical Engineering Progress, 4/13, pp. 33–40. Reproduced with permission of American Institute of Chemical Engineers.

Adsorption

In absorption, we contacted a gas with a liquid to remove or recover a component in the gas. In *adsorption*, we use a solid material to do the same thing. Adsorption can also be a liquid–solid unit operation to use a solid adsorbent to separate one liquid component from another or to remove an impurity as liquid or gas stream. The use of activated carbon to remove impurities from home drinking water is a common everyday example. Another example of this would be the use of an adsorbent to "break" an azeotrope by selectively removing one of the components via a mechanism different than vapor pressure difference. This unit operation can be used for the same purposes for which we considered absorption, but for a variety of reasons, we do not wish the recovered material to be in a solution form. The most well-known use of this technology is in the analytical chemistry area where we use the differences in adsorption preference to separate components in what is typically referred to as "gas chromatography" (GC).

This same approach can be used to separate or recover components in a gas stream on an industrial scale. Any gaseous material will have a degree of affinity for a solid surface in the same sense that it has an affinity to be absorbed into a liquid stream. We can measure this in the same sense we measure vapor–liquid equilibrium or Henry's law constants. In the area of adsorption, we call this an adsorption isotherm, and a typical example is shown in Figure 11.7, where we are plotting the amount of material adsorbed versus pressure at constant temperature.

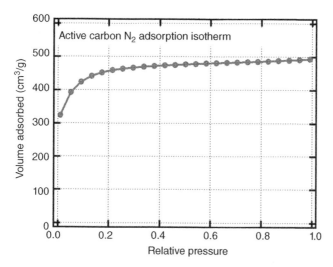

Figure 11.7 Adsorption isotherm. Source: Nandobike, https://en.wikipedia.org/wiki/File:Demac_isoth.jpg. Used under CC BY-SA 3.0 https://creativecommons.org/licenses/by-sa/3.0/. © Wikipedia.

Unless a process gas stream contains only one component that can adsorb, there will be a variety of adsorption curves, depending upon the affinity each material has for the solid adsorbent. A general equation that describes adsorption is called the Freundlich adsorption isotherm and has an equation of the following form:

$$\frac{x}{m} = Kp^{1/n}$$

where

x is the mass of the adsorbate (material adsorbed)
m is the mass of the adsorbent (material on to which the material is adsorbed)
K and n are empirical constants based on the nature of a particular adsorbent
p is the equilibrium pressure

A graphical representation of such a curve would look like that shown in Figure 11.8. This particular curve represents a material with a "K" of 4 and an "n" of 1/6.

The units in this graph are a bit different but demonstrate the same concept. "q" is in moles/kg and c is in moles/l.

A key difference between absorption and adsorption is that, in adsorption, the adsorbed material concentrates on the surface of the adsorbent as opposed to dissolving in a bulk liquid phase.

There are a number of adsorbent materials used in this unit operation, including activated carbon, zeolites, and silica gel. Their surface chemistry,

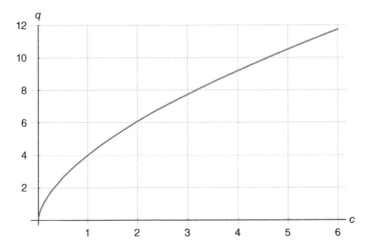

Figure 11.8 Adsorption isotherm. Source: Rosentod, https://commons.wikimedia.org/wiki/File:Freundlich_sorption_isotherm.svg. © Wikipedia.

pore size, and pore size distribution are controlled to maximize the adsorption of certain materials in preference to others. In the case of zeolites, the chemistry of preparation is so precise that the size of the pore opening is the primary controlling mechanism for what molecules are adsorbed and which ones pass through, as shown in Figure 11.9.

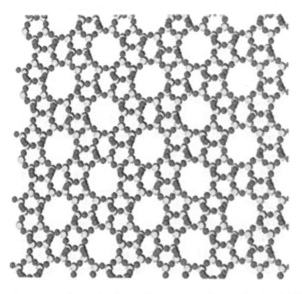

Figure 11.9 Example of a zeolite structure. Source: Prashant, http://www.nature.com/ncomms/2015/150511/ncomms8128/full/ncomms8128.html. Used under CC BY-SA 4.0 https://creativecommons.org/licenses/by/4.0/

A range of curves showing strongly adsorbed (i.e., irreversible) to non-adsorbed is shown in Figure 11.10.

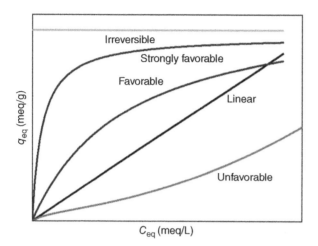

Figure 11.10 Strength of solid affinity and effect on adsorption.

In this graph, the concentration ratios are expressed in different units—another reminder how important it is to check units before comparing, sharing, or interpreting data.

We can also view adsorption as equilibrium between a gas phase component and a complex that it forms with the surface of the solid:

$$A + B \leftrightarrow AB$$

If we envision an adsorption process with a material (A) that has some reasonable ability to be adsorbed by the adsorbent (B), when the process starts, there will be no "A" leaving the column. As the adsorbent spaces are filled with adsorbate, eventually some of the adsorbate will leave the column. This can be seen in Figure 11.11.

The point at which the adsorbate begins to leave the adsorption column is called the breakthrough time and will be influenced by a number of process variables such as flow rate, pressure, temperature, and the presence of additional adsorbates. An acceptable breakthrough concentration is a function of quality specifications or environmental discharge limitations. The "AB" surface interaction mentioned previously can be complicated by the presence of more than one compound in the adsorbate as there can be additional surface chemistry interactions between the compounds on the surface after they have been adsorbed. This is why it is critical, when scaling up such a process, to use actual process stream concentrations and not single component streams, mathematically averaging the results.

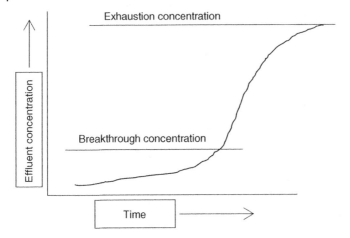

Figure 11.11 Adsorption versus time.

If we look inside the adsorbent column as a function of time, we would see something similar to that shown in Figure 11.12.

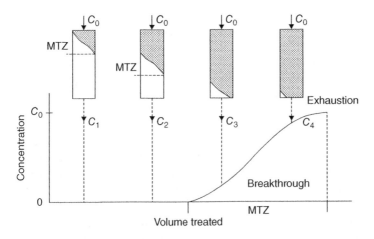

Figure 11.12 Adsorption breakthroughs.

The shape of the curve and the rate at which the adsorbate begins to leave the column will depend on the "K" and "n" of the adsorbent and the velocity of the flow.

The adsorbate (material removed from the gas or liquid) is removed from the adsorbent in a number of different ways. One is to flow an inert gas at a higher temperature than the original adsorption was done, as shown in Figure 11.13.

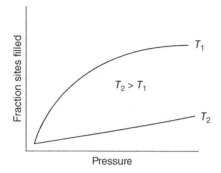

Figure 11.13 Effect of pressure and temperature on adsorption.

Another approach is to reduce the pressure and a third is to use a solvent that has a greater affinity for the adsorbate than the adsorbent.

These types of processes, since they are inherently semicontinuous, will not provide a constant flow of clean stream that contained the adsorbate. If a continuous flow of product leaving this kind of unit operation is desired, then there must be a parallel process running "out of phase," which will start as the other unit is being regenerated.

Adsorbents having a large affinity for water can be used to "break" the ethanol–water azeotrope discussed earlier by removing water and leaving pure ethanol behind.

Practical considerations in the design of such process equipment include the following:

1) Humidity of the adsorbate stream. Water can be adsorbed on material such as carbon as well as other adsorbents. If the humidity level is less than 30%, this is not normally a problem. Adsorbed water can also cause surface interactions between different adsorbates that were not planned.
2) Auxiliary process equipment such as fans and blowers will be affected by the particle size and particle size distribution of the adsorbent.

Ion Exchange

In this separation unit operation, we introduce polarity of different types and strengths to recover or remove materials from a liquid (most frequently water) and then use a desorption process similar to what we used in adsorption reversal to recover what we have removed and return the bed to its original state. As with adsorption, this is typically a semicontinuous process where a fluid stream is passed over/through an ion exchange resin bed to remove a specific ionic species and then "regenerated" to remove the adsorbed species, dispose of it, and then "recharge" the resin bed. A home water softener is the most common consumer use of such technology. A water softening resin bed removes the Ca^{++} ion in "hard" water while at the same time displacing and releasing Na^+ ion (already chemically attached on the resin surface) into the water. At some point in time, the resin becomes saturated in Ca^{++} ion, and the bed is "regenerated" by flushing a strong NaCl solution through the bed. The solution is made up indirectly by the homeowner when bags of salt are placed in a storage tank. Water is run through the salt tank, producing the salt solution. The displaced Ca^{++} ion (in the form of $CaCl_2$) is discharged into the sewer pipe of the homeowner.

Ion exchange polymer beads are typically cross-linked (to maintain some rigidity) polymers (styrene-divinylbenzene is a primary example) formed in the shape of round beads. The degree of cross-linking affects to what degree the polymer beads swell when in use and when being regenerated. Typical ion exchange resins/beads are shown in Figure 11.14.

The ionic chemistry on the surface of these beads can be positive ($Ca^{++}/K^+/(CH_3)_3NH_4^+$)) or negative ($OH^-/SO_4^{2-}/HSO_4^-/COOH^-$) depending upon the nature of the component that is being removed from the liquid stream passing through the bed. In addition to calcium, carbonates, silica, and charged organics can be removed with this type of process.

An industrial ion exchange process will have a number of tanks for processing, storage of effluent, and storage of regenerant in addition to the ion exchange unit itself.

From a chemical engineering standpoint, there are challenges in the design of such a process:

1) Pressure drop. The size and size distribution of the beads and the flow rate and liquid viscosity will all have significant effects on the pressure drop across an ion exchange bed. Ten-fold increases in viscosity can

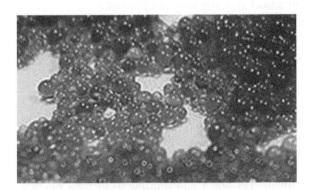

Figure 11.14 Typical ion exchange polymer bead. Source: Bugman, https://commons.wikimedia.org/wiki/File:Ion_exchange_resin_beads.jpg. © Wikipedia.

double the pressure drop through an ion exchange bed operating at 1–6 m/h linear flow rate.

2) These types of resin beads can swell as their "raw" chemistry is changed into their replaced form. This swelling can be as much as 150% by volume. It is important that sufficient free volume be available in the tanks to accommodate this swelling. If a homeowner were to look at the tank in their water softening system, they would see a very small percentage of the tank volume actually occupied by the resin, as it needs volume to expand when regenerated.

Reverse Osmosis

Reverse osmosis is a general term referring to the use of films of various porosities to allow water to flow through and reject solids, dissolved salts, ions, and biological materials such as bacteria and viruses. Osmotic pressure refers to the pressure developed across a membrane when it is placed between a salt solution and pure water, as shown in Figure 11.15.

What this means from a practical and engineering point of view is that if we want to separate water from the small molecules dissolved in it, we need to overcome this osmotic pressure in order to force water through the membrane.

This is the basis for the brackish and seawater desalinization plants around the world, which produce drinking water from water that is too salty to drink. Figure 11.16 illustrates a picture of the largest water desalinization plant in the United States using this type of technology to supply 10% of the drinking water to the city of Tampa, Florida.

Commercial membranes can be a spiral wound or of a hollow fiber type. The holes within the membranes are typically extremely small polymer tubes as seen in Figure 11.17, illustrating a seawater membrane.

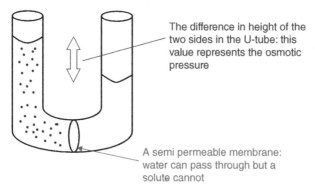

The difference in height of the two sides in the U-tube: this value represents the osmotic pressure

A semi permeable membrane: water can pass through but a solute cannot

Figure 11.15 Osmotic pressure. Source: http://creativecommons.org/licenses/by-sa/4.0/. © Wikipedia.

Figure 11.16 Tampa, FL water desalinization plant. Source: http://www.chem1.com/acad/webtext/solut/solut-4.html.

Reverse osmosis membrane element inside a pressure vessel

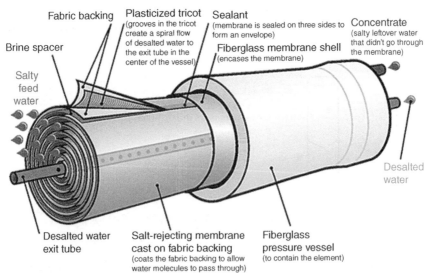

Figure 11.17 Reverse osmosis element detail. Source: http://www.chem1.com/acad/webtext/solut/solut-4.html.

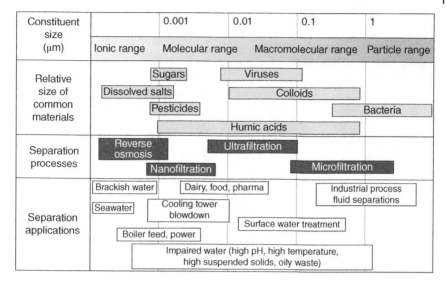

Constituent size (µm)	0.001	0.01	0.1	1
	Ionic range	Molecular range	Macromolecular range	Particle range

Figure 11.18 Differing separation capabilities of membrane systems. Source: Chemical Engineering Progress, 4/13, pp. 33–40. Reproduced with permission of American Institute of Chemical Engineers.

The ability of a given membrane design to filter and reject certain size molecules (primarily based on their molecular weight (MW) and size classifies the membranes as reverse osmosis, nanofiltration, ultrafiltration, or microfiltration) is illustrated in Figure 11.18.

Flow rates of fluids through membranes increase significantly with temperature as viscosity drops. Membrane systems must be cleaned frequently. High pressure reverse flow cleaning and chemical system cleaning are both used.

Gas Separation Membranes

Gases have different permeation rates through polymer materials and membranes. For example, most commercial food wraps sold in grocery stores are based on polyethylene. When wrapped in such a film, water does not permeate rapidly, so they can keep food moist and fresh. However some vegetable products need oxygen to avoid spoilage, and polyethylene's resistance to oxygen flow causes spoilage. A copolymer of vinyl/vinylidene chloride (commonly known as Saran™, a registered trademark of S.C. Johnson and Son, Inc.) controls both oxygen and water transmission. Some of these food wrap films are based on polyvinyl chloride polymer films.

Figure 11.19 Gas separation membrane. Source: Used with permission of Air Products.

The same kind of difference in gas permeability can be used to separate gases, as shown in Figure 11.19.

There are many situations in which it would be desirable to have an oxygen/nitrogen ratio different than the normal 79/21% (not counting rare gases). For example, in a combustion process, it might be desirable to have a higher oxygen/fuel ratio to raise the temperature of the flame. As opposed to building and operating a cryogenic air separation plant to produce liquid oxygen, a membrane that could separate the nitrogen and oxygen in air via a membrane system would have value. Nitrogen and oxygen have different permeation rates (oxygen normally higher), so in concept it should be possible to purify either one via a membrane process. In a recent example of a commercial application of this concept, a commercial gas separation membrane process has been developed that is sold to auto and tire stores to enable them more economically produce nitrogen at a given site, in order to use nitrogen as a tire filling gas versus using a compressed air process. This is shown in Figure 11.20.

This type of technology also has potential in flammability control if a way to produce nearly pure nitrogen can be developed, allowing a less expensive way to control the "oxygen" part of the fire triangle. It could also provide an enriched stream of oxygen for medical or oxidation processes vs. the purchase of liquified oxygen or oxygen cylinders.

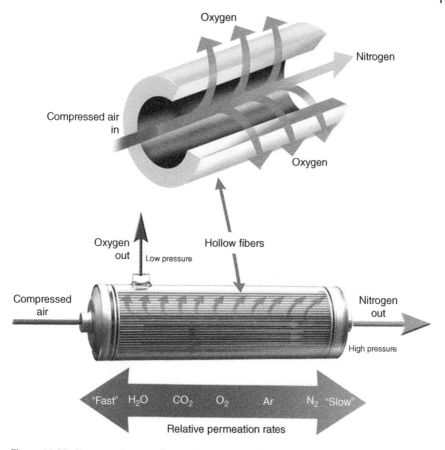

Figure 11.20 Air separation membrane. Source: Reproduced with permission of Air Products.

Leaching

A leaching operation refers to recovery of a valuable material from a solid ore or deposit through the use of a liquid, passed through the solid, and referred to as a leaching agent. Many minerals are recovered from ores in this fashion, including gold. Leaching can be done with an ore on the surface, possibly mined and brought to the surface, or "in situ" on an underground deposit.

The chemistry used in these types of operations will depend upon finding a complexing/leaching agent that can specifically attach to the mineral of interest. In the case of gold (Au is the chemical symbol for gold) mining, cyanide is such a material:

$$Au^+ + 2CN^- \rightarrow Au(CN_2)^-$$

The complexed gold is released by contacting with zinc (Zn) and displacing the leached gold:

$$2Au(CN_2)^- + Zn \rightarrow Au + Zn(CN)_4^-$$

In addition to "in-place" leaching, there are a variety of mechanical liquid–solid contacting devices used in leaching.

Chemical engineering design variables in all of these processes will include:

1) Particle size and particle size distribution of the ore from which the valuable material is being recovered. These will greatly affect the fluid flow rate.
2) Flow rate, flow distribution, density, and viscosity of the leaching fluid
3) Depth of the ore bed
4) Particle degradation during any process using active grinding equipment
5) Downstream treatment of the leached material to achieve the desired purity

The leaching of tea from a tea bag or the brewing of coffee are common household analogies to this industrial process. The particle size and particle size distribution of the tea or coffee particles inside the bag, the temperature and flow rate of hot water, and how evenly the hot water leaching liquid is distributed or contacted will all affect the taste and concentration of the resulting beverage.

Liquid–Liquid Extraction

Liquid–liquid extraction is a unit operation where a mixture of liquids, usually in one phase, is contacted with a third liquid in which one of the components in the starting material is "extracted" into the third liquid due to enhanced solubility in this third liquid. We previously mentioned the concept of azeotropes that exist in some liquid–vapor systems that limits our ability to separate the components based on vapor pressure differences. One of the alternatives to deal with this situation is to identify another liquid in which one of the components of the azeotrope is preferentially soluble, leaving behind the desired component. This requires the identification of a solvent that is insoluble in the starting mixture but has a much greater affinity for one of the components in the starting mixture. This extracted mixture must then be treated in another separate process to be able to reuse the solvent being used as an extractant. The material removed (i.e., water in the case of the ethanol/water azeotrope) must also be recovered and treated for use or disposal. These extra steps are what make liquid–liquid extraction an inherently more complex and expensive unit operation.

The solvent is the material we are adding to extract the desired or undesired material. The "extract" is the phase containing whatever material we are

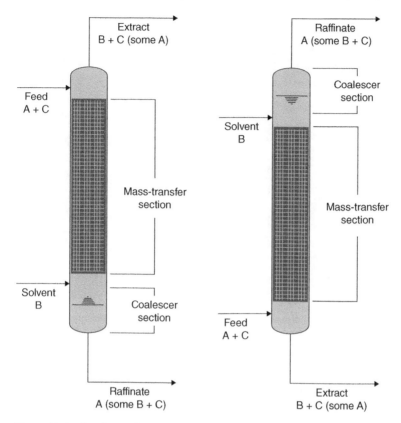

Figure 11.21 Continuous liquid–liquid extraction process. Source: Chemical Engineering Progress, 12/04, pp. 22–25. Reproduced with permission of American Institute of Chemical Engineers.

trying to extract from the initial mixture, and what is called the *raffinate* is the residual material that then must be further treated (Figure 11.22 shows the extract to be of a lower density than the raffinate phase, but this is not necessarily the case).

Liquid–liquid extraction can be done in a batch manner, but can also be done continuously as shown in Figure 11.21.

There are a number of possible configurations of such a column depending upon liquid density differences. The trays basically function as a series of tanks. The same thing could be accomplished in a series of batch tank operations.

Other approaches to continuous operations of this type include columns with a rotating shaft with the two liquids moving up and down, respectively, or columns with a vertical shaft movement as shown in Figure 11.22.

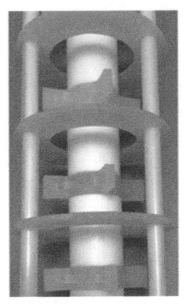

Figure 11.22 Agitated extraction columns. Source: Chemical Engineering Progress, 12/10, pp. 27–31. Reproduced with permission of American Institute of Chemical Engineers.

These types of devices usually have a variable speed drive for the internal shaft, allowing the turbulence of liquid–liquid contacting to be varied as a function of liquid physical properties such as density, viscosity, and surface tension.

Some of the process variables that can be manipulated and controlled are:

1) Ratio of solvent to feed
2) Type of physical contacting device
3) Temperatures of feed and solvent
4) Degree of agitation within the contacting column (or tank if a batch operation is used)

Surface tension, which we have not discussed previously, can affect the wetting of the process contacting surfaces as well as the tendency of the process liquids to foam when agitated. Defoaming agents may need to be evaluated and considered in such situations. Surface tensions (measured in dynes/cm) can vary from 2 to 8 for short-chain alcohols to as much as 45–50 for low MW alkanes.

The equilibrium data for a three-phase system must be available or measured in order to determine the number of contacting stages needed in a liquid–liquid extraction process. This type of data is normally shown in the form of a triangular phase diagram as shown in Figure 11.23.

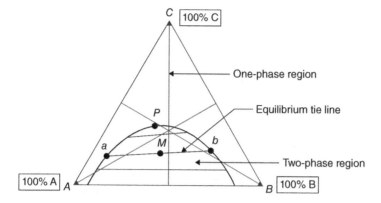

Figure 11.23 Ternary liquid equilibrium phase diagram.

In this illustration of a three-phase system, the concentrations of A, B, and C, which have been measured in laboratory studies, are shown. When the end of the line is at "A," the composition is 100% A, and it is similar with the two other components. Superimposed on this general diagram is solubility data relating to solubilities of these three liquids in each other. In the one-phase region, all three components are mutually soluble. At what is known as the "Plaitt Point" (P in the diagram), the solution "splits" into distinct phases (area labeled "M"), where we may have added "A" to the starting mixture, wanting to separate "B" from "C." If we separate these phases and add additional "A," the mixture will split again according to the tie lines in the diagram. We can envision this process as shown in Figure 11.24:

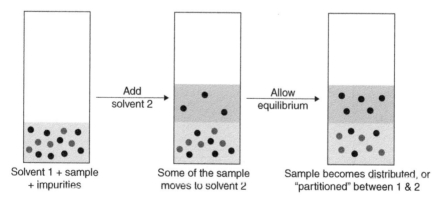

Figure 11.24 Liquid–liquid extraction in stages. Source: Reproduced with permission of American Institute of Chemical Engineers.

The composition of each phase is calculated with a mass balance calculation, summing the totals of A, B, and C.

This is analogous to the boil and condense analogy in distillation. The number of times this must be done will be a function of the solubilities of the liquid components in each other which will also be a function of temperature. A mass balance, using the compositions for each phase, will determine the final composition after each stage of contact. The number of stages of contact required to reach the final desired composition will determine the number of tanks or the height of the extraction tower.

The "tie lines" shown in Figure 11.24 are the equivalent of vapor–liquid equilibrium lines in a distillation system except that in a liquid–liquid extraction process we are viewing the data for two different *liquid* phases. The contact/separate concept here is analogous to the boil/condense model in distillation.

These kinds of diagrams are also used in solid–solid systems such as ceramics and gas–liquid–solid systems relating to systems such as the removal of caffeine from coffee beans with carbon dioxide.

It is also possible to diagram a liquid–liquid extraction process in a way similar to what we did in distillation, illustrating the analysis of how many contacting stages might be required (see Figure 11.25).

As with distillation or other mass transfer operations, the efficiency of the contacting at each stage will be a function of numerous process and physical property values (density, viscosity, and surface tension differences), and the number of stages actually required will be more than we might calculate from such a diagram.

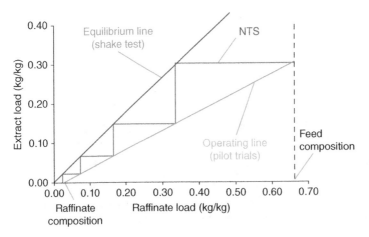

Figure 11.25 Liquid–liquid extraction stage diagram. Source: Chemical Engineering Progress, 11/15. Reproduced with permission of American Institute of Chemical Engineers.

Summary

There are a number of separation processes based on physical properties other than relative vapor pressure differences, which we used in distillation. Solubility differences can be used to recover gases into liquids. Differences in solid surface affinity and pore sizes in solids can be used to selectively recover or separate materials from liquids and gases via adsorption. Differences in liquid miscibility can be used to selectively recover or separate via liquid–liquid extraction. In many cases, several of these more specialty unit operations may be applicable to a given separation or recovery goal. The operating cost of each process, the value of the material being recovered or removed, and the capital costs of each will need to be compared to make the best choice.

Coffee Brewing
Coffee beans, either in whole or ground form, have a small percentage of actual coffee flavor ingredients with the rest being inert, as far as the coffee drinker is concerned, ingredients. In order to produce a cup of coffee, these flavor ingredients must be "leached" into water, away from the grounds. This is done by a home version of a leaching process. The beans are ground to varying degrees. A "course" grind will have a lower surface area than a "fine" grind. Everything else being equal, the fine grind will produce a stronger tasting coffee. Hot water is either percolated through or dripped through the grounds, leaching out the flavor ingredients to varying degrees. We can also assume that the geometry of the leaching mechanism and the temperature of the water (viscosity and solubility effects) will also affect the flavor of the coffee produced in this process.
The rate and temperature of the water will both affect the amount of flavor (and non-flavor) ingredients leached from the "pod" or the bed of coffee.

Discussion Questions

1 If any of these specialty unit operations are being used in your processes, are their function and design parameters well understood? What was the basis for their original choice?

2 What is the effect of changes in water temperature, viscosity, and density on their operation?

3 If less sophisticated unit operations are currently being used, is there a potential advantage in terms of cost or quality through the use of adsorption, leaching, membranes, etc.?

4 Conversely, if the choice of a specialty separation was made based on cost or lack of feasibility of a more common unit operation, have those economics changed requiring reevaluation? Could membrane quality gases replace purchased cryogenic liquids?

5 Would higher purity water have an advantage if used in your processes?

Review Questions (Answers in Appendix with Explanations)

1 Absorption is the process for recovering a gas into a:
   A __Solid
   B __Another gas
   C __Liquid
   D __Any of the above

2 Stripping is removal of a gas from:
   A __Liquid
   B __A reactor
   C __A tank truck
   D __Any of the above

3 The key variable that is used in designing an absorber or a stripper is the:
   A __Temperature of the liquid
   B __Temperature of the gas
   C __Henry's law constant
   D __External temperature

4 Henry's law constant represents:
   A __The ratio of gas partial pressure to gas concentration dissolved in the liquid
   B __The inverse of Henry's law variable
   C __The approval of Henry to the gas solubility data generated in the lab
   D __How much more gas will dissolve in a liquid if the pressure is increased

5 In an absorber, the gas enters at the:
   A __End
   B __Top
   C __Bottom
   D __Middle

6 In a stripper, the liquid enters at the:
   A __End
   B __Top
   C __Bottom
   D __Middle

7 In designing an absorber it is important to take into account:
   A __Proper distribution of inlet gas across the bottom of the tower
   B __Proper distribution of liquid over the cross sectional area of the tower
   C __Potential temperature rise due to heat of gas dissolution
   D __All of the above

8 Demisters may be required at the top of a stripper due to:
   A __Fill in void space
   B __Control of possible liquid carryover
   C __Operators are sad when seeing material being removed from a liquid
   D __To supply pressure drop

9 Adsorption is the process for recovering a component from a fluid or gas onto a:
   A __Solid
   B __Membrane
   C __Liquid
   D __Any of the above

10 The efficiency of adsorption is governed by:
   A __What kind of carbon is used
   B __Affinity of the gas for the solid
   C __Pore size of the adsorbent
   D __All of the above

11 Variables that can affect the efficiency and selectivity of adsorption include:
   A __Temperature
   B __Pressure
   C __Adsorption isotherms
   D __All of the above

12 Adsorption beds can be regenerated by all of these techniques except:
   A __Change in pressure
   B __Change in temperature
   C __Seriously wishing
   D __Purging with a large amount of gas to displace the adsorbed material

13   Liquid chromatography is a unit operation that utilizes_____to recover and/or separate liquid components:
   A __Molecular size
   B __Surface charge
   C __Liquid–solid surface chemistry
   D __Any of the above

14   Ion exchange processes use what functionality bound to a polymer surface to achieve separation:
   A __Ionic charge
   B __Pore size
   C __Differing molecular weight polymer additions
   D __Surface roughness

15   Ion exchange beds are regenerated through the use of:
   A __Change in pressure
   B __Change in temperature
   C __Large volumes of the opposite original charge solutions
   D __Purging with a large amount of gas to displace the exchanged material

16   A serious practical issue when regenerating ion exchange beds is:
   A __Using the wrong regenerant solution
   B __Hydraulic expansion
   C __Noise created
   D __Regenerating the wrong bed

17   Liquid–liquid extraction is a unit operation involving the use of a material's preference to be dissolved in:
   A __One liquid close to its boiling point versus another liquid at room temperature
   B __One liquid close to its freezing point versus another liquid at room temperature
   C __One liquid close to its critical point versus another liquid near its boiling point
   D __One liquid versus another liquid

18   To design a liquid–liquid extraction process, the following is needed:
   A __A ternary phase diagram
   B __Knowledge of densities and density differences
   C __Surface tension of process fluids
   D __All of the above

19 Operating and design variables for a liquid–liquid extraction operation include:
  A __Temperature
  B __Contact time
  C __Liquid physical properties
  D __All of the above

20 Leaching is a unit operation used to:
  A __Go back to the days of the gold rush
  B __Recover money from a stingy relative
  C __Recover a material from a solid via liquid contact
  D __Recover a material from a solid via gas contact

21 Membranes separate materials based on the difference in their:
  A __Molecular weight and size
  B __Desire to go through a very small hole
  C __Value and price
  D __Cost

Additional Resources

Chen, W.; Parma, F.; Patkar, A.; Elkin, A., and Sen, S. (2004) "Selecting Membrane Filtration Systems" *Chemical Engineering Progress*, 12, pp. 22–25.

Dream, B. (2006) "Liquid Chromatography Process Design" *Chemical Engineering Progress*, 7, pp. 16–17.

Ettouney, H.; El-Desoukey, H.; Fabish, R., and Gowin, P. (2002) "Evaluating the Economics of Desalination" *Chemical Engineering Progress*, 98, pp. 32–39.

Fraud, N.; Gottschalk, U., and Stedim, S. (2010) "Using Chromatography to Separate Complex Mixtures" *Chemical Engineering Progress*, 12, pp. 27–31.

Harrison, R. (2014) "Bioseparation Basics" *Chemical Engineering Progress*, 10, pp. 36–42.

Ivanova, S. and Lewis, R. (2012) "Producing Nitrogen via Pressure Swing Adsorption" *Chemical Engineering Progress*, 6, pp. 38–42.

Kenig, E. and Seferlis, B. (2009) "Modeling Reactive Absorption" *Chemical Engineering Progress*, 1, pp. 65–73.

Kister, H. (2006) "Acid Gas Absorption" *Chemical Engineering Progress*, 6, pp. 16–17.

Kucera, J. (2008) "Understanding RO Membrane Performance" *Chemical Engineering Progress*, 5, pp. 30–34.

Shelley, S. (2009) "Capturing CO_2: Gas Membrane Systems Move Forward" *Chemical Engineering Progress*, 4, pp. 42–47.

Teletzke, E. and Bickham, C. (2014) "Troubleshoot Acid Gas Removal Systems" *Chemical Engineering Progress*, 7, pp. 47–52.

Wang, H. and Zhou, H. (2013) "Understand the Basics of Membrane Filtration" *Chemical Engineering Progress*, 4, pp. 33–40.

Video: https://www.youtube.com/watch?v=KvGJoLIp3Sk (accessed on September 22, 2016).

12

Evaporation and Crystallization

Let's now look at two unit operations that are used to do two things:

1) Increase the concentration of a solution containing one or more dissolved materials. A simple example would be a salt solution (NaCl, KCl, $CaCl_2$, etc.). Though the most common use of this unit operation is in concentrating salt solutions, the principles are equally applicable to concentrating a material in an organic solvent, such as a drug from an organic solvent.

2) Precipitate a material, which has value, from a solution. It is to be noted that evaporation can do this as well, but evaporation is the term usually used to describe a process with heat input and a crystallization process used to describe one that is using cooling. Again, this can refer to either a water-based solution or an organic solvent-based system.

Evaporation

In evaporation, we start with a solution of one or more solid materials, which are dissolved in a solvent. A common example would be salt in water. It could also refer to a pharmaceutical product dissolved in an organic solvent. This unit operation refers to the vaporization of the solvent and increasing the concentration of the "solute" (the material dissolved in the solvent). Evaporation, carried to its extreme, would leave only a solid behind. Usually the valuable material is what is dissolved in solution.

Evaporation is a specialized form of heat transfer, which we discussed in Chapter 8. We supply steam, or some other hot heating medium, to a heat exchanger, and this heat input boils the solvent, increasing the concentration of the solute. Since the steam is condensing and the solution is boiling under turbulence on the other side of the heat exchanger, heat transfer coefficients tend to be higher than those seen in conventional liquid–liquid shell and tube heat exchangers. The mechanical design of such a heat exchanger can be of several different types.

Chemical Engineering for Non-Chemical Engineers, First Edition. Jack Hipple.
© 2017 American Institute of Chemical Engineers, Inc. Published 2017 by John Wiley & Sons, Inc.

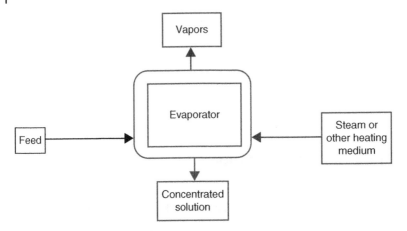

Figure 12.1 Evaporation process.

Horizontal or vertical tubes, inserted into process vessels, with the vapor coming off the top of the evaporator are condensed or disposed of. There are many different mechanical configurations of these types of evaporators depending upon liquid properties, type of heat source being used, and nature of the final product. A general diagram of a simple evaporation process is shown in Figure 12.1.

The details of the evaporator design can vary greatly, depending upon the nature (density, viscosity) of the feed, its boiling point versus pressure, the concentration desired, and the sensitivity of the concentrate to higher temperature. This last point is very important in the food and drug industry.

Regardless of the details, the overall design of the evaporator (it is just another form of a heat exchanger we discussed previously) will follow the general overall heat exchange formula we reviewed before:

$$Q = UA\Delta T$$

Q will be the total amount of energy required (BTU/h. or kcal/h.) to boil a certain number of pounds of water or solvent times its heat of vaporization plus any sensible heat necessary to reach the solution's boiling point plus any external heat losses from the process vessels. A will be the heat transfer surface area available, which could be a combination of heat transfer coils inserted into the evaporator and heat transfer area on the jacket of the evaporator. As we discussed earlier, U is the overall heat transfer coefficient and in a turbulent boiling situation will typically be higher than a normal liquid–liquid or liquid–gas heat exchanger.

There are three practical issues that often cause an evaporator to not perform as designed:

1) Level control in the evaporator must be sufficient to cover all of the tubes (assuming that all the tube area was used in the design). We have not discussed process control up to this point, but suffice it to say that if the level in the evaporator is not high enough to cover all of the tubes, the performance will not be what is expected.
2) Fouling can reduce the heat transfer coefficient over time. This would normally be seen as a slow decline in performance over time, and some degree of overdesign in the area that would be used to compensate for this. The comments made earlier in the chapter on heat exchanger apply equally to evaporators.
3) Boiling point rise. This is a subtle design mistake that is possible to make. In the basic equation mentioned earlier, the ΔT should be the difference between the temperature of the heating medium (steam, hot oil, etc.) and the temperature of the *product* solution at its boiling point. If there is a salt solution that is being evaporated, its boiling point will rise with a rise in concentration. (This would not usually be the case with an organic solvent being evaporated.) If the boiling point is underestimated, the ΔT will be lower and thus the driving force for heat transfer less, requiring more heat transfer area to achieve the desired result.

This rise in boiling point with salt solutions is characterized by what is known as a Duhring plot, where the boiling point of a particular solution is shown graphically against the boiling point of water. These diagrams can be readily found in the literature.

Operational Issues with Evaporators

Some practical issues in the operation of evaporators include the following:

1) Steam Condensate Removal. The steam typically used in an evaporator condenses into hot condensate and is either returned to an internal power plant via a steam trap system or is disposed of to public waterways, usually after having been cooled in a pond or a cooling tower. If the path for steam condensate flow is blocked in any way, hot liquid condensate can build up in the steam chest of the evaporator, reducing the heat transfer coefficient. Steam–liquid heat transfer is greater than liquid–liquid heat transfer.
2) Buildup of Non-condensables. If the steam being used is coming from a water supply that has not been vacuum degassed, there is the possibility that very small amounts of inerts in the boiler feedwater (nitrogen, oxygen) will

very slowly build up in the steam chest of an evaporator. Over a period of time, this will reduce the partial pressure of the steam and lower the heat transfer rate.

3) Entrainment and Foaming. Salt solutions tend to have high surface tension values, making it possible to create stable foams and froth during boiling. This problem can be addressed in a number of ways. First, antifoam agents can be added to the solution, assuming that the addition of these materials does not interfere with quality specifications or downstream use. Second, the vent line leaving the evaporator can be greatly enlarged in diameter (compared to the evaporator itself) to lower the gas velocity leaving the evaporator, making coalescence and condensing back into the evaporator easier. Another approach sometimes used is to use a direct steam nozzle into the overhead vapors, accelerating coalescence of drops into a liquid stream that will drain back into the evaporator.

We have discussed physical properties several times already, and this is another area where we need to remember these fundamentals and make sure we know the data for the materials we are handling. As salt solution (calcium chloride in the graph) concentration rises in an evaporator, its viscosity will increase as shown in Figure 12.2.

In Chapter 8, we discussed the heat transfer coefficient being proportional to the Reynolds number to the 0.8 power. Since the Reynolds number is $DV\rho/\mu$

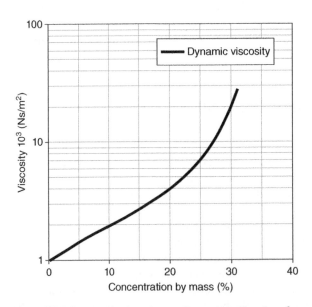

Figure 12.2 Increase in viscosity as salt concentration rises. Source: Reproduced with permission of Engineering Toolbox.

(μ is viscosity), the Reynolds number will decrease as the viscosity increases. We need to make sure when designing the evaporator that we are using the actual physical properties inside the evaporator and not the feed going into the evaporator. Viscosities typically increase as salt concentration increases.

There a number of different evaporator configurations used in different industries based on the unique characteristics of the materials being concentrated. This is especially true in the food industry.

Vacuum and Multi-effect Evaporators

Evaporators do not need to be run at atmospheric pressure. If we raise the pressure, the boiling point of the solution will rise; if we increase the pressure of steam feeding an evaporator the ΔT between the heating medium and the boiling point will increase. If we run the evaporation under vacuum, the boiling point of the solution will drop below its value at atmospheric pressure, again increasing the ΔT and the rate of heat transfer. The choice between atmospheric and non-atmospheric pressures for evaporation will depend upon a number of factors. Again, in many areas of chemical engineering, there is no right or wrong, but choices.

1) In general, vacuum is a more energy intensive process than pressure. Vacuum can be produced by a steam-driven vacuum jet or a mechanical compressor. Steam jets, though having no moving parts, produce a waste water stream, which must be treated and/or disposed of.
2) Evaporation with higher pressure steam will not only increase the ΔT but will also increase the capital cost of the evaporator.
3) Using high pressure steam in combination with vacuum on the solution side will not only provide the highest ΔT but will also require the highest capital with the lowest steam use. There is a trade-off between cost and capital, as is usually the case in all "optimization" of chemical engineering unit operations.

A multi-effect evaporator is one that uses the evaporated water from the first stage to provide the energy input to the second stage in the following fashion, showing a two-stage multi-effect evaporation system (Figure 12.3).

The feed (1 in the diagram) is heated and evaporated (in heat exchanger A1 with steam entering at point 3). The output from this first step enters a separation tank (B1). The hot, partially concentrated solution is sent to the second stage, while the vapor from the first stage (leaving the top of B1) is used to further evaporate the solution. This can only be accomplished if the pressure is reduced to lower the boiling point. The final product (2) leaves the bottom of the second stage (B2), and #4 indicates the flow of vapor to a vacuum system, which has created the vacuum necessary.

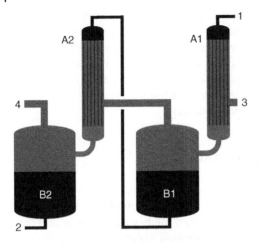

Figure 12.3 Multi-effect evaporator. Source: https://commons.wikimedia.org/wiki/File:Double_effect_evaporator.PNG. Used under CC BY-SA 3.0, https://commons.wikimedia.org/w/index.php?curid=214928. © Wikipedia.

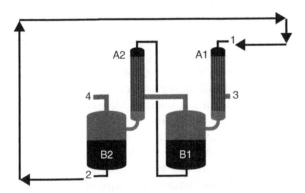

Figure 12.4 Use of vapor recompression in evaporation. Source: CC BY-SA 3.0, https://commons.wikimedia.org/w/index.php?curid=214928

Another option used when electrical costs are relatively low compared with capital costs is vacuum recompression, where we take the off-gas from the evaporator under vacuum and then recompress it to be the primary energy source. This would effectively take the steam leaving the second stage (#4) and reusing it as the feed to the first stage, as shown in Figure 12.4.

Multiple effect evaporation can also be used to produce drinking water from seawater, as opposed to the membrane separation processes discussed earlier. The choice is a function of energy and capital costs.

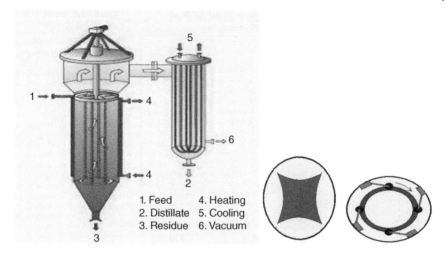

1. Feed
2. Distillate
3. Residue
4. Heating
5. Cooling
6. Vacuum

Figure 12.5 Wiped film evaporator mechanisms for viscous solutions. Source: Reproduced with permission of Sulzer.

If the viscosity of the fluid being concentrated increases rapidly during the evaporation process, it may be necessary, in a tube-type evaporator, to wipe the tubes continuously to prevent plugging of the evaporator tubes. Figure 12.5 shows an illustration of this type of evaporator/concentrator as well as some of the mechanical design choices used in scraping the walls as the evaporation takes place.

As we have discussed previously, there is no one right answer to an evaporator design choice as there was no one answer to a distillation column design. The costs of capital and energy will be key decision input data along with physical building restrictions and of course the physical properties of the solution being evaporated or concentrated.

Crystallization

In crystallization, we are normally referring to precipitating a solid from a solution by lowering its temperature. Since most materials' (but there are exceptions!) solubility in water, or other solvents, increases with temperature, one way of recovering a solid from a solution is to cool it and some of the solids will drop out of solution ("precipitate") to follow a solubility curve. Such curves for several common salts are shown in Figure 12.6.

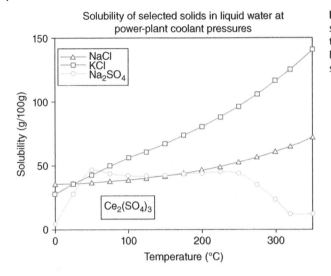

Solubility of selected solids in liquid water at power-plant coolant pressures

Figure 12.6 Selective salt solubilities versus temperature. Source: From public literature sources.

There are several important things to notice in this graph:

1) Two of the salts (NaCl, (common table salt); and KCl) consistently increase in solubility (shown in g/100 g in this figure) as the temperature is raised, but the slopes of these curves vary significantly. Potassium chloride (KCl) increases in solubility at a far greater rate than ordinary table salt (NaCl) as the temperature is raised.

2) Sodium sulfate's (Na_2SO_4) solubility also increases up to about 60°C, then levels out, and finally *decreases* in solubility beyond 220°C. This is due, in part, to the interaction of the sulfate molecules with the water solvent in a way that would not happen with the chloride salts. We would not necessarily expect to see this same kind of deviation in a nonpolar solvent, and it would be a great mistake to make that assumption.

3) Cesium sulfate ($Ce_2(SO_4)_3$) and other salts in the calcium family *decrease* in solubility as temperature is raised. This behavior is the root cause of hard water deposition on clothes being washed in hot water.

One approach to crystallization is to use what is known as *evaporative crystallization*. In this case, we evaporate the solution to the point where solids precipitate out when their solubility limits have been exceeded.

This type of crystallizer is similar to an evaporator, except that the solution is boiled beyond the solubility point and a solid slurry product is produced. This wet slurry then needs to be processed further.

Crystallization is inherently a more complicated process than evaporation as the usual focus is not a more concentrated solution, but recovery of a component in the liquid phase that has value. Part of this value may be in the shape, size, particle size, and particle size distribution of the crystals. Crystallizers are usually agitated, in order to better control particle size and particle size distribution, adding another process variable. Another factor is that crystal shape and particle size distribution are usually part of the design criteria in a crystallizer. Many salts, especially simple inorganic chemical salts, can have multiple states in conjunction with waters of hydration, each of which has a different crystal shape. Crystallizers are frequently used to produce drugs and food products. The shape, average particle size, and particle size distribution of these products are critical to their biological function in terms of dissolution and uptake rates.

Considering the details of an agitated crystallizer, there are a number of mechanisms going on in parallel:

1) Precipitation of material
2) Growing the size of a crystal through deposition of new material on the surface of previously precipitated material
3) Existing crystals being broken up by the agitator
4) Impurities or crystallization solution (frequently referred to as "mother liquor") being trapped inside the precipitating and growing crystals

Crystallization processes can be run under many different conditions:

1) Batch, vacuum. The feed solution is loaded into a vessel followed by pulling vacuum on the vessel, reducing the boiling point of the solution. The solution boils and material precipitates out, producing slurry for further processing, as shown in Figure 12.7. "Mother liquor" is a term used to describe the feed solution.

Figure 12.7 Batch vacuum crystallizer. Source: Chemical Engineering Progress, 10/08, pp. 33–39c. Reproduced with permission of American Institute of Chemical Engineers.

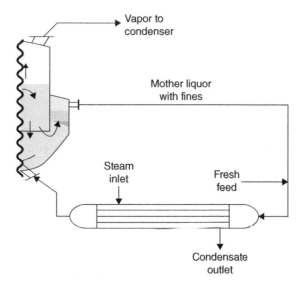

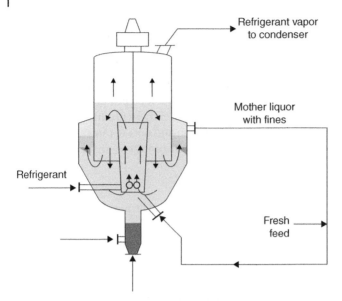

Figure 12.8 Forced circulation with indirect cooling. Source: Chemical Engineering Progress, 10/08, pp. 33–39c. Reproduced with permission of American Institute of Chemical Engineers.

2) Forced Circulation with Indirect Cooling. This process can be continuous, and it is critical to ensure that the temperature drop across the heat exchanger is small to minimize the potential for solids to cake and plug up the heat exchanger (Figure 12.8).

In addition to the particle size and entrained mother liquor in the crystals, there are a number of other product characteristics affected by the design and operation of a crystallizer:

1) Agglomeration and Caking. The nature of the particle size and distribution will have an effect on the material's cohesive bonding and strength, which will in turn affect to what degree a solid agglomerates and cakes in downstream storage in silos, bins, and hoppers as well as in shipping containers.
2) Surface area, which will impact the items mentioned in (1), can also impact the interaction of the crystalline product with other materials, especially if it is used in a catalyst formulation.
3) Morphology. This refers to the shape and structure of the crystal. Anyone who has shopped for a diamond understands that the shape and structure of the same carat weight diamond, and how it refracts light, can have a significant impact on its perceived value and price. The same can be true for specialty chemicals.
4) Bulk Density. The particle size and particle size distribution can greatly affect how much volume a given amount of weight of material occupies. This will affect packaging size and the design of bulk storage equipment.

5) Bioavailability. If the crystalline product is used in a pharmaceutical or drug application, the rate of dissolution in the stomach and/or blood stream can be critical issues. Let's assume that we have a distribution of particle size in a drug that looks like that in Figure 12.9.

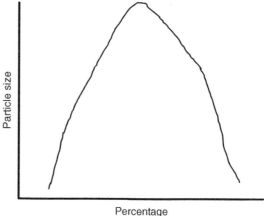

Figure 12.9 Particle size distribution.

If small particle size dissolves faster than a large particle size, then this crystal will most of its material dissolved after being taken or absorbed, with a small fraction dissolving quickly and another small fraction dissolving quickly at the end. If we wanted to have the crystalline drug dissolve all at once, very quickly, we would design the crystallization process to produce a narrow range of very small particles, which when compared with the previous particle size distribution, would look like that shown in Figure 12.10.

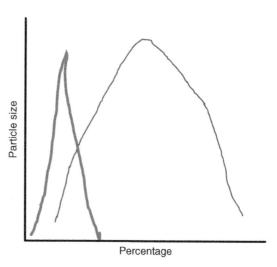

Figure 12.10 Changing particle size distribution.

This type of particle size distribution is the basis for such products as rapidly dissolving aspirin to deal with heart attacks.

It is also worth noting that whatever is done to create a certain particle size, shape, and distribution will have a major effect on downstream processing including drying, conveying and transport, and storage. We will discuss these unit operations in Chapters 13 and 14.

Crystal Phase Diagrams

Most of the time when we think about salts being soluble or insoluble, we tend to think in terms of "in solution" or "out of solution." In many cases, especially in the inorganic salt world, things are a bit more complicated. Many inorganic salts can form *hydrates* with water. For example, calcium chloride ($CaCl_2$) can exist in its normal anhydrous (without water) form or it can exist as a hydrated crystal, that is, $CaCl_2 \cdot 2H_2O$. This material, if we go back to the periodic table and calculate the weights of the various molecules, is approximately $111/147$ or 75.5% $CaCl_2$. Though this material is approximately 25% water, it looks like a white solid. Calcium chloride also forms hydrates with 1 and 6 moles of waters of hydration. Sodium chloride (NaCl) on the other hand has no hydrates and is either in solution or not. A crystalline phase diagram shows where and which hydrates form at various temperatures and concentrations. This information is critically important to have when operating an evaporator or crystallizer as it clearly defines what kinds of materials it is possible to have at any condition.

One of the most interesting and complex examples of such an inorganic salt phase diagram is that for magnesium sulfate ($MgSO_4$). This salt forms hydrates with 0.5, 1, 2, 4, 6, and 7 moles of hydration. The compound with 7 moles of water, $MgSO_4 \cdot 7H_2O$, is the product sold in grocery and drug stores as "Epsom salts." Again, if we calculate the percentage of $MgSO_4$ in the product, it is approximately $120/120 + 126$ or 49% $MgSO_4$. But if we pour it out of its container, it appears to be a normal anhydrous solid. Figure 12.11 shows the phase diagram for this compound and its waters of hydration.

There are some very important points that we know about this system by simply looking at the phase diagram:

1) A *liquid* solution of $MgSO_4$ can only exist at conditions indicated on the left side of the phase diagram (approximately 0–40%) $MgSO_4$ over the temperature range of 25–200°F.
2) At each weight fraction and temperature, the phase diagram tells us which species will exist; a mass balance calculation will tell us how much of each we have. For example, between 120 and 150°F and a concentration

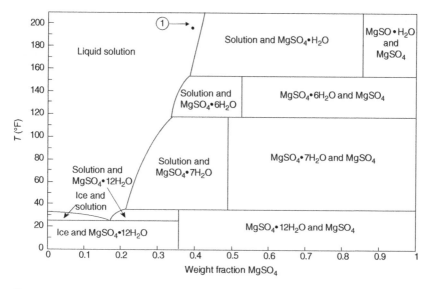

Figure 12.11 Phase diagram for magnesium sulfate ($MgSO_4$). Source: Reproduced with permission of NASA.

(weight fraction) of 53–100% $MgSO_4$, we will have a mixture of two solids, $MgSO_4 \cdot 6H_2O$ and $MgSO_4$ (anhydrous). We do not get to choose what we have within this part of the diagram. That is why it is so important to know whether a salt being crystallized has such a diagram and what it looks like.

3) The consequence of not understanding such a diagram (if it exists) is obtaining an undesired product and, from a practical and operating perspective, plugged lines and vessels when solids are present or when liquids were assumed to be present.

Supersaturation

This is a process technique occasionally used to produce very fine particle size. All of us have heard the tale about the frog sitting in a pan of hot water, and the temperature being raised so slowly that the frog never realizes what is happening and gets scalded to death before it realizes how hot the water is. The reverse of this is true for crystallizers. It is possible to cool a saturated salt solution extremely slowly so that, in effect, the solution forgets that it is supposed to precipitate out the salt it contains. At some point though, the phase diagram takes over and a large amount of very fine crystals is generated very rapidly.

Crystal Purity and Particle Size Control

As the crystals are precipitating from the mother liquor, there will be some entrapment of the solvent (from which they are precipitating from) inside the crystals. This will be affected by the rate of crystallization, the particle size distribution, particle morphology, and surface tension effects. Depending upon the quality requirements for the final product, it may be necessary to redissolve the product from the first crystallization and recrystallize (possibly in a slightly different manner) to produce the product quality needed.

Particle size control will be affected by all the process variables we have discussed, and again there may need to be second or third crystallization to obtain the crystal purity and particle size distribution required. It is critical to understand the customer and business requirements as crystallizers are designed and operated.

Summary

The unit operations of evaporation and crystallization are used to concentrate liquid solutions, with dissolved solids, or to precipitate and recover materials dissolved in solution as products. Optimization of these processes depends upon a thorough understanding of the physical chemistry of the solutions (including phase diagrams and liquid properties such as density, surface tension, and viscosity), the particle size and particle size distribution requirement of the final products, and how the wet solids will be further processed. These downstream processes, which we will discuss in Chapters 13–15, include filtration, drying, and solids handling and storage.

Evaporation in Coffee Brewing

As the coffee from a drip coffeemaker drops into a carafe that sits on a hot plate, evaporation starts, and the rate is determined by the basic equations discussed in this unit. The higher the temperature of the hot plate, the higher the rate of evaporation, significantly increasing the evaporation of water and the concentration of the remaining coffee. As discussed earlier, this will also increase the kinetic rate constant of the coffee degradation chemistry. There are coffee carafes that have no hot plates and are some form of a vacuum bottle. Since there is no way for water vapor to escape, evaporation does not occur but flavor and taste (chemical) degradation still continue, but at a lower rate.

If a traditional coffee carafe on a hot plate is left long enough, solids will begin to precipitate out of solution, just as in a crystallizer.

Discussion Questions

1 If your processes use evaporation, what is the desired concentration? How is it controlled? How is its operation affected by changes in steam pressure or temperature of the heat source?

2 How is the level in the evaporator controlled? Is it ensured that all of the steam tube area is used?

3 If a salt solution is being evaporated, is it known whether there is a phase diagram? If so, what is the impact of operating changes?

4 What can affect the heat transfer coefficient inside the evaporator?

5 To what degree is fouling an issue?

6 If crystallization is being used, is the phase diagram understood (if applicable)?

7 How do changes in operating conditions (agitation speeds, rate of temperature decrease, etc.) affect particle size and particle size distribution? What is the effect on any changes on product use and quality?

Review Questions (Answers in Appendix with Explanations)

1 Evaporation involves concentrating:
   A __A liquid in a solid
   B __A solid in a liquid
   C __A gas in a liquid
   D __A liquid in a gas

2 The primary design equation for an evaporator considers:
   A __Boiling point of the liquid
   B __Pressure in the evaporator
   C __Temperature difference between the heat source and the boiling point of the solution
   D __All of the above

3 The boiling point of a solution to be evaporated with steam is affected by all but:

A __Pressure

B __Steam price

C __Concentration of dissolved salts

D __Temperature differential between heat source and boiling point of the solution

4 The boiling point of a salt solution will ____ with increased concentration of the dissolved salt:

A __Decrease

B __Need more information to answer

C __Increase

D __Rise by the square root of concentration change

5 If the steam pressure feeding an evaporator slowly decreases with time, the salt concentration leaving the evaporator will ____over the same time period:

A __Increase

B __Decrease

C __Stay the same

D __Depends (on what?) ____

6 Salt solution carry over into the vapor phase of an evaporator can be minimized through the use of:

A __Prayer

B __Filters

C __Cyclones

D __Demisters

7 Multi-effect evaporators function by:

A __Using the vapor from one stage to vaporize another stage

B __Using the "super-effect" of steam

C __Condensing the first stage vapor and then boiling it a second time

D __Taking advantage of off-peak power prices

8 Film evaporators are used primarily for:

A __Temperature-sensitive and high viscosity materials

B __Emotionally sensitive materials

C __Hold your temper materials

D __All of the above

9 The basic difference between evaporation and crystallization is that the solution is concentrated by:

A __Use of diamonds

B __Cooling

C __Any type of heat sources that is available except steam

D __Suction

10 The types of crystals produced in a crystallizer are affected by:

A __Phase diagram

B __Rate of cooling

C __Amount of agitation

D __All of the above

11 A phase diagram for a salt and solvent will determine all but:

A __The types of crystals that will be formed

B __Where various hydrates will form as a function of temperature and concentration

C __Cost incurred to operate at a particular point within the phase diagram

D __How to produce certain types of salt hydrates

Additional Resources

Genck, W. (2003) "Optimizing Crystallizer Scaleup" *Chemical Engineering Progress*, 6, pp. 26–34.

Genck, W. (2004) "Guidelines for Crystallizer Selection and Operation" *Chemical Engineering Progress*, 10, pp. 26–32.

Glover, W. (2004) "Selecting Evaporators for Process Applications" *Chemical Engineering Progress*, 12, pp. 26–33.

Panagiotou, T. and Fisher, R. (2008) "Form Nanoparticles by Controlled Crystallization" *Chemical Engineering Progress*, 10, pp. 33–39.

Samant, K. and O'Young, L. (2006) "Understanding Crystallizers and Crystallization" *Chemical Engineering Progress*, October, pp. 28–37.

Wibowo, C. (2011) "Developing Crystallization Processes" *Chemical Engineering Progress*, March, pp. 21–31.

Wibowo, C. (2014) "Solid-Liquid Equilibrium: The Foundation of Crystallization Process Design" *Chemical Engineering Progress*, March, pp. 37–45.

13

Liquid–Solids Separation

In Chapter 12, we discussed the unit operations of evaporation and crystallization, both of which produce slurries or moist products (moist in this context could be water or some other solvent). It is now necessary to recover and dry the product for final storage and use.

Filtration and Filters

If an evaporation or crystallization process has produced wet slurry, we now need to filter it to produce a nearly dry cake, prior to drying. It is also possible that, in filtration, we are looking at a process to remove solid particles from a liquid product stream but the principles are the same.

Filtration is defined as the separation of solid particles from a liquid stream by forcing the slurry through a filter medium that allows the liquid to pass through, leaving the solids behind. This unit operation requires a pressure differential to force the fluid through the medium. The pressure differential can be either positive or negative, meaning that the upstream pressure can be above atmospheric pressure, or negative, meaning that a vacuum is pulled on the downstream side of the filter, "sucking" the fluid through the medium.

A simple diagram of a filter might look like that shown in Figure 13.1.

Materials that are too large to pass through the holes in the filter media are retained and build up into a filter cake and those smaller pass through into and with the filtrate.

Chemical Engineering for Non-Chemical Engineers, First Edition. Jack Hipple.
© 2017 American Institute of Chemical Engineers, Inc. Published 2017 by John Wiley & Sons, Inc.

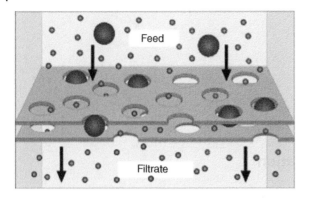

Figure 13.1 Basics of filtration. Source: Wikiwayman, https://commons.wikimedia.org/wiki/File:FilterDiagram.svg. Used under CC BY-SA 3.0, https://creativecommons.org/licenses/by-sa/3.0/deed.en. © Wikipedia.

Filtration Rates

The rate at which liquid flows through the solids will depend on a number of factors:

1) Pressure differential across the medium (similar to the discussion in Chapter 7 on fluid flow)
2) The liquid properties such as density, viscosity, and surface tension
3) Particle size and particle size distribution of the solids being filtered (this would be similar to the difference we see in the rate of coffee production through coarse vs. espresso coffee grounds)

Many types of slurries have particle sizes small enough to clog the pores of a particular filter medium, thus "blinding" it off. In these cases a filter aid is preloaded on to the filter medium to provide a barrier between the fine particles and the filter cloth or medium. Inert materials frequently used for this purpose include sawdust, carbon, diatomaceous earth, or other large particle size materials.

Filtration can be a batch or continuous unit operation. If done continuously, there must be a mechanism to constantly remove some of the filtered material. If batch, the filtration is continued until the space between the filter leaves is full or the pressure differential possible with the process is reached. In any design, there must be a pressure differential to make the fluid pass through the solids. This pressure differential can be produced by pumps on the upstream side of the filter, vacuum on the downstream side, or some combination of both as a function of time.

Filtration Equipment

One of the most commonly used continuous filtration processes is what is known as *rotary vacuum* filtration, as illustrated in Figure 13.2.

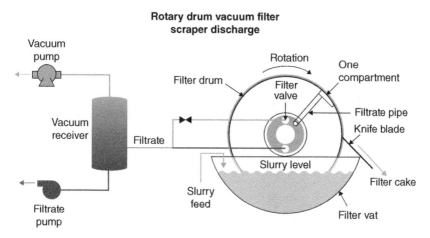

Figure 13.2 Rotary vacuum filter. Source: Reproduced with permission of Komline.

In this process, liquid is fed into a feed vat and "sucked" up on to the filter cloth, which rotates on a rotating drum. The vacuum is provided through a sealed system within the drum and connected to an external vacuum receiver and pump. As the cake travels around the drum (rotating clockwise), it is "dewatered" and then as it reaches the other side of the drum, the cake is scraped off as a filter cake and sent on for further processing (drying and storage), as shown in Figure 13.3.

In the mineral industry, filtration is used to recover mining cake and it is often sufficiently dry so as not to require a knife blade or other removal device—a simple sharp turn of the belt causes the cake to drop off the filter media. Another approach occasionally used for cake removal is a rotating brush system.

In addition to these types of continuous belt filters, there are other types of commercial filters. One is a semicontinuous filter known as a plate and frame filter. This used filter cloths wrapped around a frame with space between to serve as the reservoir to hold the solids. When the spaces are filled, the filter is shut down and the collected solids are removed, usually by a pressure source, into a collection pit, as shown in Figure 13.4.

If the fluid being filtered is the product of interest and the objective is to remove large-sized contaminants, gravity filtration, possibly supplemented by vibration, may also be used.

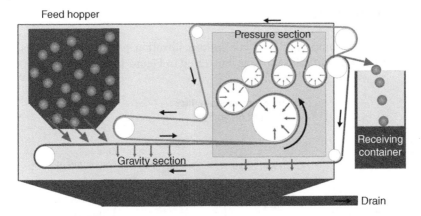

Figure 13.3 Belt filter. Source: Wikiwayman, https://commons.wikimedia.org/wiki/
File:BeltPress.svg. Used under CC BY-SA 3.0, https://creativecommons.org/licenses/by-
sa/3.0/deed.en. © Wikipedia.

Figure 13.4 Plate and frame filter. Source: Wikipedia Public Domain License, http://fr.
wikipedia.org/wiki/Utilisateur:Roumpf

Some of the design and operational parameters of filters and the symbols
used in this area are the following:

V refers to the volume of filtrate.
P or ΔP refers to the pressure or pressure differential across the filter and its
medium.
ω (omega) is used to refer to the weight of solids per unit volume ($\#/\text{ft}^3$, kg/m^3).
α (alpha) is used to describe the *compressibility* of the filtered cake. The higher
the α, the more compressible is the cake, meaning that as the pressure is
increased on the filter cake, the void volume decreases, slowing the filter
rate. You can consider sand and rubber at opposite extremes of this value.

Figure 13.5 Filtration rate versus cake compressibility.

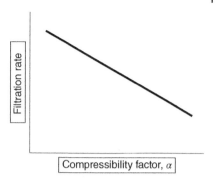

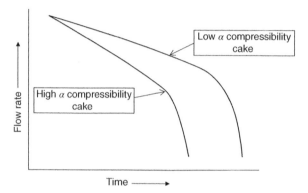

Figure 13.6 Flow rate at constant pressure with high and low compressibility cakes.

If we were to plot filter rate as a function of compressibility, we would see a graph as in Figure 13.5.

There are two basic ways of running a filtration operation—constant pressure or constant rate, though it is possible to combine one with the other depending upon practical limitations.

If we run the filtration at constant pressure, the filter cake will build up over time, providing additional resistance to the flow of fluid. The degree to which this happens will depend, in part, on the compressibility of the cake. A flow versus time diagram would look like as shown in Figure 13.6. In this figure is also displayed what the flow rate versus time curve would be seen with a more compressible (higher α) cake.

A more compressible cake, with a higher α, will tend to fill in the voids in the filtered solids more rapidly as the filtration progresses. At some point in time, the flow of filtrate becomes too low and uneconomical, and the filter is shut down, the cake removed, the filter medium washed, and the process restarted. The washing process and its fluid rates will be also affected by the physical properties of the solids recovered.

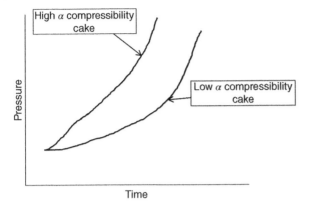

Figure 13.7 Pressure requirement versus time for constant rate filtration and high and low compressibility cakes.

If we desire to have constant flow, then the pressure must increase with time to overcome the pressure drop increase as the filter cake increases in thickness over time, as shown in Figure 13.7.

The general design equation for a filtration integrates the flow, pressure, solids content, and cake compressibility as follows:

$$V = \frac{PA^2\theta}{\alpha w}$$

where

V = volume of filtrate
P = system pressure
A = area of filter medium
θ = time
α = cake compressibility
w = solids concentration in mass/unit volume (i.e., #/ft^3)

If we increase the pressure, area, or length of time, we will get increased filtration rate. If we increase the cake compressibility or the solids content in the slurry being filtered, the filtration rate will drop.

In most cases, the solids recovered on a filter media will need to be washed to remove the last quantities of mother liquor from the original crystallization. In rare cases, where extreme purity is needed, the solids may be redissolved, recrystallized, and refiltered. As previously mentioned, the principles of design and performance would be the same.

Centrifuges

All of us have seen and used a centrifuge device in our daily lives—our home washing machines. Centrifuges are filters with the addition of centrifugal force to increase the pressure drop across the filter and increase the flow rate and/or amount of fluid passed through.

Figure 13.8 shows an illustration of a commercial continuous centrifugal separator being used to separate oversize particles from a slurry. The degree of separation can be controlled by design variables such as rotational speed and the choice of internal screens. It is also important to remember that any mechanical device of this kind will affect the particle size and particle size distribution of filtered solids due to the mechanical and abrasive forces acting on the solids while being separated.

The efficiency or added force of a rotational centrifuge is proportional to its radius and angular velocity squared, or $r\omega^2$. They can be batch (as in a home washing machine) or continuous; if a mechanism for continuous removal of the filtered cake is provided (in Figure 13.8), an internal screw conveying device is used to accomplish this. The spinning pushes the mother liquor through the filter medium, and in a semicontinuous fashion, a "pusher" moves through the middle of the machine, pushing the solids out to a temporary storage area for additional processing such as drying, particle size reduction, or agglomeration.

Operational and design issues for these types of devices include the following:

1) Specific gravity differences between the liquid and solids being filtered
2) Solids concentration in the feed
3) Liquid throughput rate
4) Rotational speed
5) Degradation of particle size due to the mechanical contact and pushing (not a concern with simple filters)

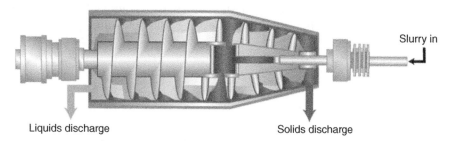

Liquids discharge   Solids discharge

Figure 13.8 Decanting centrifuge. Source: Chemical Engineering Progress, July 2012, pp. 45–50. Reproduced with permission of American Institute of Chemical Engineers.

6) Particle shape and compressibility
7) Rinsing and washing capabilities
8) Density and viscosity properties of the liquid phase
9) Efficiency of solids capture
10) Liquid clarity leaving the centrifuge

A summary of the various types of centrifuges and their capabilities in different aspects is shown in Table 13.1.

Table 13.2 shows another view of the various performance aspects of different types of centrifuges.

This type of equipment will invariably involve tests with numerous vendors who specialize in each of these types of equipment.

Particle Size and Particle Size Distribution

Regardless of what kind of filter or centrifuge is used, the particle size and particle size distribution will affect its performance. Smaller particles will have a tendency to blind off the filter media, and the acceleration rate of smaller versus larger particles in a centrifuge will produce a nonuniform cake composition (in terms of particle size). This difference may need to be taken into account in downstream processing in drying and solids handling.

Liquid Properties

Again, regardless of the type of filter being used, the properties of the liquid, from which the solids are being filtered, will greatly affect the flow of liquid through the filter. The effects of liquid properties will be similar to those discussed earlier in the unit on fluid handling, especially if a centrifuge is used. A high viscosity liquid will flow through the solids and filter media slower; a high density fluid will flow through faster. If we have an unusual fluid that has response to shear, its viscosity will be affected by the rotational speed of the centrifuge.

Summary

Liquid–solids separation is an important unit operation in chemical engineering. The ability of a filter to separate a solid from a liquid depends on the physical properties of the liquid and solid being filtered, the pressure drop across the filter, the nature of the solids in terms of particle size, particle size distribution, and compressibility.

Table 13.1 Centrifuge choices.

| | Vertical basket | | | Horizontal peeler | Inverting filter with PAC | Decanter |
	Manual discharge	Peeler discharge	cGMP			
Diameter, mm	200–1600	800–1800	Up to 1250	250–2000	300–1300	~1500
Operation	Batch	Batch	Batch	Batch	Batch	Continuous
Cake washing	Yes	Yes	Yes	Yes	Yes	No
Discharge	Manual	Automatic	Automatic	Automatic	Automatic	Automatic
Containment	No	Yes	Yes	Yes	Yes	Yes
Solids filterability	Low to medium	Low to medium	—	Medium	Medium	—
Volume capacity	Low	Medium to high	—	High	Medium to high	High

Source: Chemical Engineering Progress, July 2012, pp. 45–50. Reproduced with permission of American Institute of Chemical Engineers.

Table 13.2 Comparison of various types of centrifuges.

Performance parameter	Vertical basket	Horizontal peeler	Inverting bag	Screen bowl	Scroll screen	Pusher	Vibratory	Imperforate basket	Tubular	Solid bowl	Disc
Particle size, μm	10–100	10–100	10–100	50–5000	100–50,000	15–100	500–10,000	1.0–1,000	0.01–100	5–5,000	0.1–100
Wt.% solids in feed	5–50	5–50	5–50	5–40	30–80	30–60	40–80	1–50	<0.1	0.5–60	0.1–5
Liquid flow rate, gal/min	1–150 (feed)	1–150 (feed)	1–50 (feed)	1–500	5–200	5–250	300–1,000	1–100	<25	1–600	2–600
Solids production rate, ton/year	0.01–1.5	0.01–1.5	0.01–0.5	1–100	40–350	5–50	5–50	0.01–3	<0.003	0.05–100	0.001–1.5
Centrifugal force, m/s^2	500–1500	500–1500	500–1500	500–2000	500	500	500–2,000	1,000–20,000	4,000–10,000	1,000–10,000	5,000–10,000
Solids wash	E	E	E	F	F–G	F–G	P	NA	NA	P	NA
Solids capture	E	E	E	G	G	G	G	E	E	E	E
Liquid clarity	G–E	G–E	G–E	F–G	F–G	F–G	F	E	E	E	E

Source: Chemical Engineering Progress, August 2004, pp. 34–39. Reproduced with permission of American Institute of Chemical Engineers.
E, excellent; G, good; NA, not applicable; P, poor.

Back to Coffee Brewing
The ground coffee that is used in a coffee brewing machine will have different particle size and different particle size distribution depending on the grinding setting at the store or in last minute grinding at home. These grounds, of whatever average size and size distribution, are placed in a filter. These filters are of various shapes and geometries. The flow rate through the coffee filter along with the grind nature will affect the "strength" of the coffee in terms of its taste. Percolator coffee brewers, which use a recycling of filtrate through the ground beans, will use very coarse grinds, while espresso coffee machines will use extremely fine particles to produce a very strong coffee flavor with a once through brewing. The temperature of the water used will affect not only the temperature of the final cup of coffee but also its viscosity and how fast it flows though the grounds. The final step in making coffee is a chemical engineering filtration process.

Discussion Questions

1 If you are running filters or centrifuges, is the basic information about flow rates versus pressure known?

2 If a filter with a pre-coat is used, are the limitations of its particle size limitations known?

3 Depending on whether constant pressure or rate filtration is being used, how was that decision reached? Should it be revisited?

4 Should the choice of continuous versus batch be revisited?

5 If a centrifuge is being used, what was the basis for its design, type, and vendor made? How reliable is its operation?

6 Would a different particle size distribution feeding either a filter or centrifuge improve their operation? If so, how could this be achieved?

Review Questions (Answers in Appendix with Explanations)

1 The driving force for filtration is:
   A __Pressure differential
   B __Concentration differential

C __Temperature differential

D __Temperament differential

2 A pre-coat on a filter medium may be required if:

A __It is cold in the filter operations room

B __The operating instructions say so

C __The particle size of the solids is greater than the hole size in the filter medium

D __The particle size of the solids is smaller than the hole size in the filter medium

3 If the filtration is run under constant pressure, the flow rate will ____ with time:

A __Drop

B __Stay the same

C __Increase

D __Need more information to know

4 If the filtration is run to produce constant volume output, the pressure will ____ with time:

A __Drop

B __Stay the same

C __Increase

D __Rise by the cube of the change in flow

5 Raising the solids concentration in a filter feed, with other variables unchanged, will ____ the filtration rate:

A __Increase

B __Decrease

C __Not affect

D __Drop by the square root of the solids concentration change

6 Increased compressibility of a filtration cake will ____ the filtration rate over time:

A __Increase

B __Decrease

C __Stay the same

D __Increase by the square of the compressibility

7 A centrifuge adds what force to enhance filtration rate:

A __Gravity

B __Pressure

C __Centrifugal/centripetal

D __Desire for a faster rate

8 The rate of filtration in a centrifuge is proportional to the ____of the rotational speed:

A __Linearity
B __Square root
C __Cube
D __Square

Additional Resources

Norton, V. and Wilkie, W. (2004) "Clarifying Centrifuge Operation and Selection" *Chemical Engineering Progress*, August, pp. 34–39.

Patnaik, T. (2012) "Solid-Liquid Separation: A Guide to Centrifuge Selection" *Chemical Engineering Progress*, 108, pp. 45–50.

14

Drying

On the assumption that the material that has been filtered is the product of interest or if the "wet" product has come from some other part of a process or unit operation, it usually needs to be dried, meaning that the residual solvent or water needs to be evaporated to some degree. To what degree is primarily a function of customer requirements, but to some extent also the nature of the physical handling of the product after drying. An example of this latter case might be caking of the product while being stored or in shipping containers.

In order for a solvent or water to evaporate, energy must be supplied from some source and a driving force must exist. It normally comes from indirect heating with steam but could come from hot oil. A vacuum can be added to the process to increase the vapor pressure difference and accelerate the drying process. The drying can also be accomplished by direct contact with a hot air or gas stream, which then may need to be treated in some way to remove any entrained solid particles.

Since drying may involve a temperature increase in the solid material, we must also check for any possibility of product decomposition (from a quality standpoint) or possibly decomposition that may lead to a safety or reactive chemical incident.

We can view the general nature of a drying process in Figure 14.1. This graph shows what we would find if we ran a drying test on the wet solid.

At the start of a drying process, there is usually substantial moisture (or solvent) content on the surface of the solid. In this early stage, there is little mass transfer resistance of the moisture evaporating. It is based almost solely on the contacting between the solid and the drying gas.

As the drying process progresses, the moisture (or solvent) needs to diffuse to the surface of the solid to be able to contact the drying gas. This point, where the rate of drying begins to decline significantly, is referred to as the "critical moisture content" or CMC. The drying rate continues to

Chemical Engineering for Non-Chemical Engineers, First Edition. Jack Hipple.
© 2017 American Institute of Chemical Engineers, Inc. Published 2017 by John Wiley & Sons, Inc.

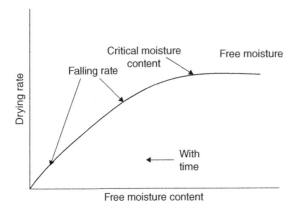

Figure 14.1 Drying rate versus residual moisture content.

fall as the residual moisture/solvent begins to need to overcome strong capillary surface forces within the pores of the solid. During these latter phases, the time it takes to remove the water/solvent increases significantly, and the economic value of achieving these low values must be weighed against the cost of the extended drying time and additional energy cost.

In general, the rate of drying will be proportional to the temperature differential between the drying heat source and the temperature of the material being dried as well as the area of the material being exposed to the drying medium and inversely proportional to the heat of vaporization. The contact area can be increased by agitation, tumbling within batch dryers, and baffling inside rotating equipment.

There are a number of commercial type dryers, including:

Rotary Dryers

These dryers can be either continuous or batch and can be direct heat fired or the water/solvent evaporated with the assistance of vacuum. Batch processing would be typically used in specialty and pharmaceutical applications where lot control is important.

A continuous countercurrent dryer would tumble the solids within an inclined rotary cylinder with a hot gas flowing countercurrently, with possible internal baffles to minimize sticking to the walls and maximize gas–solid contacting (Figure 14.2).

Figure 14.2 Rotary dryers. Source: Brookoffice, https://commons.wikimedia.org/wiki/
File:Single_Shell_Rotary_Drum_Dryer.jpg. Used under CC BY-SA 3.0, http://
creativecommons.org/licenses/by-sa/3.0. © Wikipedia.

Spray Dryers

In this dryer configuration a concentrated slurry or solution is passed through a
spray nozzle, with high pressure drop. This atomizes the solution, creating a
very large surface area and rapid drying rates. The drying medium is typically
hot air or, in the case of some oxygen-sensitive products, a hot inert gas such as
nitrogen (Figure 14.3).

Figure 14.3 Schematic of spray drying.
Source: https://commons.wikimedia.org/wiki/
File:Spray_Dryer.gif. Used under CC BY-SA 3.0,
http://creativecommons.org/licenses/by-
sa/3.0. © Wikipedia.

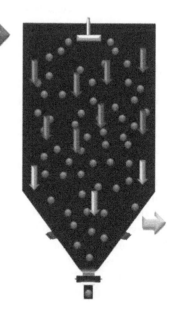

Figure 14.4 Spray nozzle for dryer. Source: Chemical Engineering Progress, 12/05. Reproduced with permission of American Institute of Chemical Engineers.

Since this dryer is contacting slurry with a fast velocity gas, which may cause some particle degradation, it is normal to see a cyclone and possibly a bag filter used to ensure that small particle solids are not discharged to the atmosphere.

The dryer requires a rather sophisticated spray nozzle to cause atomization of the slurry. Figure 14.4 shows an illustration of different discharge patterns from such a nozzle that will affect the drop size and drop size distribution. The patterns and liquid particle size distribution will also be affected by the liquid properties as well as the use of air to increase the degree of atomization.

If we recall some of the variables we discussed in the fluid handling unit, you can understand that pressure drop across the nozzle, density, viscosity, and spray stream particle size will have a great effect on the performance of such dryers.

Fluid Bed Dryers

The term "fluid bed" refers to a situation where the velocity of a gas stream is sufficient to suspend a solid. If this is done correctly, a great deal of turbulence and contact between the gas and slurry is generated, producing very rapid drying rates. The solid, if viewed from above, appears to be a suspended liquid. Any solid particle has a velocity at which it will be suspended by a gas flow. This will be a function of particle size and density, as well as gas properties. This is illustrated in Figure 14.5.

If the gas velocity is sufficient to suspend the wet solids, the drying rates will be very fast and the residence time will be low. This may be beneficial if short residence time is necessary due to temperature sensitivity. The downside of these dryers is that the turbulence will typically cause degradation in particle size and the need to use cyclones and scrubbers on the outlet gas stream to prevent significant discharge to the atmosphere.

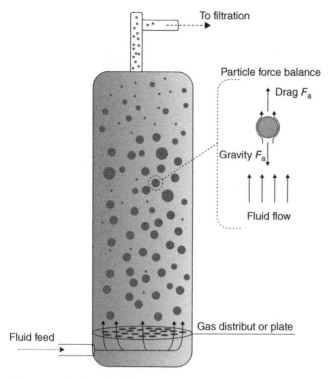

Figure 14.5 Fluidized bed particle forces. Source: https://commons.wikimedia.org/w/index.php?curid=3982684. Used under CC BY-SA 4.0, https://creativecommons.org/licenses/by-sa/4.0/. © Wikipedia.

The wet slurry and hot gas enter through support plates with a velocity sufficient to suspend the solids. These types of dryers can also be designed so that the solids move from left to right across a decreasing flow and temperature of gas and finally discharging. The gas velocity must be balanced against the force required to lift the solids out of the dryer completely. Support and distribution plates, similar in concept to those used in packed distillation towers, are used to distribute the drying gas evenly as well as to prevent solids from leaving the bottom of the column.

Belt Dryer

This is a dryer where the wet solid is deposited on a moving belt and a slow velocity hot gas is moved countercurrently, with possible staged temperature control, as shown in Figure 14.6 (drying pasta).

Figure 14.6 Food belt dryer. Source: https://commons.wikimedia.org/wiki/File:DEMACO_
DTC-1000_Treatment_Center_for_Fresh_Pasta_Production_(April_1995)_003_crop.jpg.
Used under CC BY-SA 4.0, https://creativecommons.org/licenses/by-sa/4.0/deed.en. ©
Wikipedia.

Freeze Dyers

When we need to dry very heat-sensitive materials (e.g., fruits and vegetables), it is not possible to expose them to any significant amount of heat. Most of the time, water is the material that needs to be removed and water (and other materials) has an interesting phenomenon known as a "triple point" in its solid–liquid–gas phase diagram as illustrated in Figure 14.7.

In the case of water, this "triple point" occurs around 0°C and 4 mmHg absolute pressure. Though it is expensive to produce vacuum, this is not so low as not to be economical for valuable food products such as coffee ("freeze dried"), nuts, fruits, and vegetables. It is a batch process where a high volume of product is put into a vacuum chamber and then left for a predetermined amount of time, the vacuum released, and the product then packaged. The type of equipment used in this drying process is usually a simple tray dryer with plates or shelves inserted into a chamber, the chamber sealed, vacuum created and held for a specific time, and then the vacuum released and product removed and packaged.

Summary

Drying is a unit operation used to remove water or solvents from products prior to final handling and storage. A drying curve is generated for any given product to develop a rough estimate of the cost and time involved in achieving

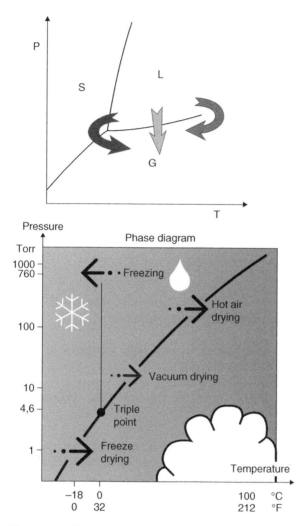

Figure 14.7 Phase diagrams with triple points. S, solid; L, liquid; and G, gas. Source: Used under https://commons.wikimedia.org/wiki/File:Drying.svg. © Wikipedia.

a given residual level. There are many different types of commercial equipment used, both batch and continuous. Vendors are normally heavily involved in specifying equipment through trials with customers, as much of the design data is empirically based. The degree of "dryness" or residual water or solvent will have a significant impact on further downstream processing such as solids handling.

Coffee Brewing and Drying

We normally don't think about drying associated with coffee brewing. We don't normally specify the moisture content of the coffee (in whatever form) being used. If it were an important variable, the way the coffee or its grounds are stored will affect the amount of water lost in storage. To a very minor degree, the degree of residual water content will have a minor impact on the concentration of the brewed coffee. One of the first commercial uses of freeze drying was in the manufacture of freeze-dried coffee. This drying and evaporation process, at low temperature and high vacuum, allows moisture to be removed without subjecting the coffee to high temperatures, thus minimizing the taste degradation of the coffee. Since this drying process is far more energy intensive than normal drying and evaporation, the cost of such coffee is generally higher.

Discussion Questions

1   Are the drying rate curves known for all the products being dried? What could affect these? What process conditions could affect?

2   Is your drying process robust enough to deal with variations in both customer requirements and changes in utilities?

3   How were the current drying processes chosen? Have alternatives constantly been reviewed?

4   How does your drying process affect particle size and particle size distribution? How does this affect your internal solids handling and your customer's use of your product?

5   If your dryer was capable of producing an extremely dry product, would that provide additional market opportunities?

Review Questions (Answers in Appendix with Explanations)

1   Drying is defined as the removal of ___ from a solid material:
    A __Solvent
    B __Coolant
    C __Water
    D __Spirits

2   Drying rate is affected by all but:
   A __Solvent concentration at any point in time
   B __Cost of vacuum or steam
   C __Agitation within the dryer
   D __Temperature difference between solid and heating medium

3   Key variables in the design and operation of a spray dryer include:
   A __Liquid or slurry to gas ratio
   B __Viscosity of fluid and pressure drop across the spray nozzle
   C __Temperature difference between hot drying gas and liquid
   D __All of the above

4   Design issues with rotary dryers include:
   A __Possible need for dust recovery
   B __Particle size degradation
   C __Dust fires and explosions
   D __All of the above

5   Freeze drying is a potential practical drying process if:
   A __A freezer is available
   B __The S–L–V phase diagram allows direct sublimation at a reasonable vacuum
   C __It is desired to have a cold product
   D __The plant manager owns stock in a freeze dryer manufacturing company

6   A drying rate curve tells us:
   A __How fast a solid will dry
   B __How much it will cost to dry a solid to a particular residual water or solvent level
   C __The drying rate curve as a function of residual water or solvent
   D __How the cost of drying is affected by the rate of inflation

7   Auxiliary equipment frequently needed for a drying process include:
   A __Cyclones and scrubbers
   B __Backup feed supply
   C __Customer to purchase product
   D __Method of measuring the supply chain

Additional Resources

Heywood, N. and Alderman, N. (2003) "Developments in Slurry Pipeline Technology" *Chemical Engineering Progress*, 4, pp. 36–43.

Langrish, T. (2009) "Applying Mass and Energy Balances to Spray Drying" *Chemical Engineering Progress*, 12, pp. 30–34.

Moyers, C. (2002) "Evaluating Dryers for New Services" *Chemical Engineering Progress*, 12, pp. 51–56.

Purutyan, H.; Carson, J., and Troxel, T. (2004) "Improving Solids Handling During Drying" *Chemical Engineering Progress*, 11, pp. 26–30.

15

Solids Handling

Safety and General Operational Concerns

As was the case with filtration and drying, this area of chemical engineering is rarely taught as part of chemical engineering curricula, and as a result, most of the science and design in solids handling is in the knowledge base and experience of vendors and engineering specialists within large corporations who handle solids as a significant part of their business. Since there is little basic training in this area, there are some sad side effects that are seen in industry:

1) Start-up times in plants handling solids are significantly greater than those handling just liquids and gases.
2) After start-up is over, the final operational conditions may vary significantly from the intended design.
3) Serious and catastrophic dust explosions and fires result from the lack of knowledge of the flammability and explosion hazards of solids and dusts. These fires and explosions occur most frequently where solids are concentrated or energy input is significant, such as in hoppers and silos, grinders and pulverizers, conveying systems, and mixing/blending equipment.

We tend to see most dust explosions in the food, wood, chemical, metal, rubber, and plastics industries where the basic materials have some natural flammability.

One of the important differences between solids and their bulk liquid and gas counterparts is the fact that solids can have different particle sizes (liquids can be in this area as well if there is an atomization or spraying process involved) when they have significant differences in surface area and energy. Here are some examples of materials and natural surface area differences:

Chemical Engineering for Non-Chemical Engineers, First Edition. Jack Hipple.
© 2017 American Institute of Chemical Engineers, Inc. Published 2017 by John Wiley & Sons, Inc.

Table 15.1 Particle size illustrations.

Beach sand	100–10000 μm
Fertilizer, limestone	10–1000
Fly ash	1–1000
Human hair	40–300
Cement dust	4–300
Coal dust, milled flour	1–100
Smoke from synthetics	1–50
Iron dust	4–20
Smoke	0.01–0.1
Paint pigments	0.1–5
Smoke, natural materials	0.01–0.1

Table 15.2 Strength of flame front for solids.

St 0:		*0*	Silica
St 1:	0–200 bar-m/s	Weak	Milk, zinc, sulfur, sugar, chocolate
St 2:	200–300	*Strong*	Cellulose, wood, polymethylmethacrylate
St 3:	>300	*Very strong*	Al, Mg, anthraquinone

Solids have been classified into general hazard classifications, similar to what is done with liquids and their flash points:

Even a material as mundane as sugar has significant fire and explosion potential, as seen in the Chemical Safety Board video in 2006, reviewing the dust explosion disaster at Imperial Sugar. A fine sugar dust explosion can generate a pressure in excess of 100 psi in less than 100 ms and can generate a flame velocity of over 500 ft/s. The lower limit for flammability of sugar (analogous to the lower explosive limit (LEL) of a flammable liquid) is approximately 9%, far lesser than the normal oxygen content of air.

The assessment of potential fire and explosion hazards with solids and dusts is a bit more complicated than the fire triangle discussed in Chapter 2. We normally represent it in the diagram shown in Figure 15.1.

We need not only the fuel (the solid), oxygen, and a source of ignition but also a way of either suspending the solid dust (increasing its surface area) or concentrating it by confinement.

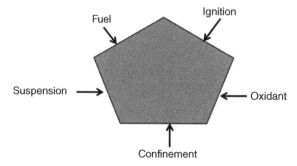

Figure 15.1 Requirements for a dust explosion.

Prevention of dust explosions concentrates in these following areas:

1) Understanding the flammability of limits of the solids being handled and how they are affected by particle size and particle size distribution
2) Minimizing points where solids can concentrate and build up, such as joints
3) Proper use of explosion relief devices and proper vent sizing
4) Pressure sensors within solids transport systems
5) Necessary grounding
6) Inerting with noncombustible gases when necessary and checking for integrity of vacuum seals
7) Use of nonconductive coatings

There are a number of tests that are used to evaluate the various aspects of solids safety:

Explosion severity test measures the maximum pressure generated by a solids explosion. This is somewhat analogous to the maximum explosive pressure discussed earlier for liquids.

Minimum ignition energy (MIE) is identical to the same type of value for liquids and gases. There will always be some amount of energy necessary to initiate a fire or explosion.

Minimum autoignition temperature (MAIT) of dust cloud is again analogous to the same value for a liquid or gas.

Minimum explosive concentration (MEC) of dust in air is analogous to the LEL for a liquid or gas in air. Below some level of fuel (solid), there is insufficient fuel to sustain the fire.

Limiting oxygen concentration (LOC) is again identical to the same measurement obtained for liquids. There is a minimum oxygen content in the gas phase, which is required for a solids fire to sustain itself.

Electrostatic chargeability test (ECT) provides a measure of the ability of a solid to hold a charge, which then may serve as an ignition source later on.

Solids Transport

Solids need to be moved for various reasons:

1) Unloading of solid raw materials to be used in reactions
2) Transfer of solids from a dryer into a solids classification, size reduction unit operation, or directly into storage
3) Loading of outgoing trucks or barges

Solids transport is done in many different ways:

1) Screw conveyors
2) Bucket elevators
3) Belt conveyors
4) Pneumatic conveyors

Screw conveyors are tubular devices with an internal rotating helical screw (referred to as a "flighting") that moves the solid from one place to another. A general diagram of screw conveyor is shown in Figure 15.2.

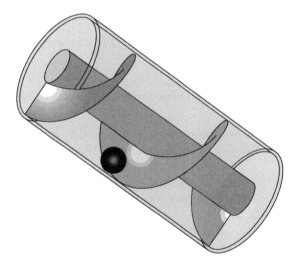

Figure 15.2 Screw conveyor. Source: Silberwolf, https://commons.wikimedia.org/wiki/
File:Archimedes-screw_one-screw-threads_with-ball_3D-view_animated_small.gif. Used under
CC BY-SA 2.5 https://creativecommons.org/licenses/by-sa/2.5/deed.en. © Wikipedia.

The primary design variables are the diameter and depth of the trough, the size and rotational speed of the screw, the angle or pitch of the screw, the depth of solid during transport, and the clearance between the screw and the wall. If the solid has significant dust explosion potential, there also may be safety relief, inerting, and monitoring systems.

Another view of an operating screw conveyor is shown in Figure 15.3.

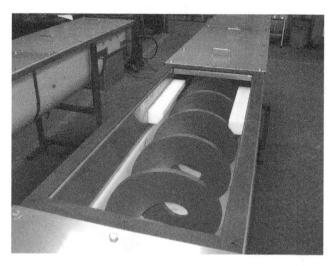

Figure 15.3 Operating screw conveyor. Source: https://commons.wikimedia.org/w/index. php?curid=9515471. Used under CC BY 3.0 https://creativecommons.org/licenses/by/3.0/ deed.en. © Wikipedia.

Design variables would include the following:

1) Speed of the transfer screw.
2) Wall clearance between the screw and the wall. Close clearance will ensure not only more uniform directional flow but will also cause wear and potential corrosion products entering the product being transferred.
3) "Starved" or full-fed transport. This will affect the rate of transport, the degree of abrasion, and the degree of particle size attrition.
4) Need for cooling or heating on the jacket of the screw conveyor.
5) Screw conveyors, due to the abrasive nature of the transport mechanism, can also provide some mixing within them as well as reduction in particle size while being conveyed.

When a screw conveyor is being used to control and feed material into a reaction or mixing system in a controllable fashion, it is often referred to as a screw meter.

The energy consumption of a screw conveyor will depend on a number of system properties, including auger and screw diameter as shown in Figure 15.4.

Costs will also be affected by the solids density (the higher, the more energy needed), the RPM of the screw (the faster, the more energy consumed, and the greater the particle size attrition), and the inclination angle.

Bucket elevators are used to transport solids vertically as shown in Figure 15.5.

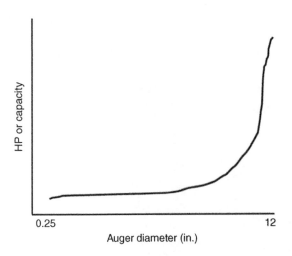

Figure 15.4 Screw conveyor energy costs.

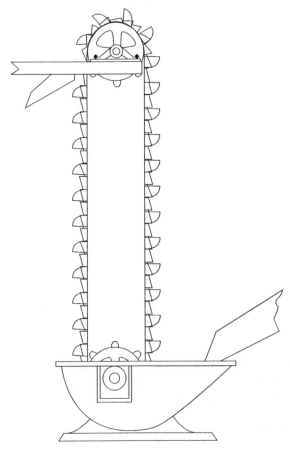

Figure 15.5 Bucket elevator. Source: Henry Kreitzer Benson, https://commons.wikimedia.org/wiki/File:Bucket_elevator_drawing.JPG. © Wikipedia.

Figure 15.6 Mineral belt conveyor.

Belt conveyors, frequently used in large-volume transport in the mining and mineral industry, are illustrated in Figure 15.7.

This conveyor shown in Figure 15.6 is carrying coal, while an illustration of the one carrying sulfur from a mining operation on to a ship is shown in Figure 15.7.

Design variables will include speed, width, and energy consumption. The general design equation for a belt conveyor would be expressed as $Q = \rho AV$, where Q is the amount of material transported, ρ is the density of the material, A is the cross-sectional area of the solid on the belt, and V is the velocity of the belt.

Pneumatic Conveyors

Given a sufficient amount of gas, a solid particle can be "lifted" and transported along with the gas. This is called pneumatic conveying. Figure 15.8 shows a generic flow diagram of such a system, along with its auxiliary support equipment.

Figure 15.7 Belt conveyor for sulfur. Source: Leonard G, https://commons.wikimedia.org/wiki/File:AlbertaSulfurAtVancouverBC.jpg. Used under CC SA 1.0 https://creativecommons.org/licenses/sa/1.0/. © Wikipedia.

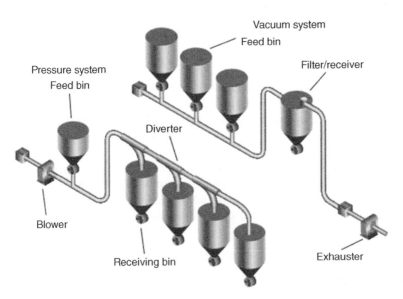

Figure 15.8 Pneumatic conveying systems. Source: Chemical Engineering Progress, 12/05, pp. 22–30. Reproduced with permission of American Institute of Chemical Engineers.

Material is typically fed into a storage bin via one of the many solids transport devices discussed earlier. In this case, a screw conveyor is being used. Material from a bin is combined with an airstream (this stream could be an inert gas such as nitrogen if there was a flammability concern) that combines with the solid and is conveyed, with sufficient gas velocity, into a receiving bin. This bin could be used for final storage or for unloading into hopper car. As we can see, there are numerous controls on this system to measure level, gas flow, and so on.

Some of the general design considerations and concerns are as follows:

1) Vacuum or Pressure Driven. The gas stream needs a pressure differential to flow. This can take the form of a positive pressure on the upstream side or vacuum/suction on the receiving side. There are pros and cons to both. A pressure-driven system may cause leaks of solids and dust to the atmosphere. A vacuum system may allow air leakage in, which may cause oxidation of the solid or create a hazardous atmosphere. In general, for the same pressure drop, the vacuum is more expensive than the pressure-driven system.

2) Particle Size Distribution. If the solids being transported have a wide range of particle size, the smaller particles will tend to travel faster, and the solids distribution in the receiving vessel will be different than that of the original storage vessel. This will need to be taken into account when designing the receiving vessel to make sure adequate flow of solids out of the bottom of the receiving vessel is unhindered.

3) Abrasion. If solids are transported at a high rate in a gas stream, we may need to worry about abrasion to the transporting pipes. This will depend upon the natural abrasiveness of the solids, the velocity of the gas stream, and the material used for the transport pipe.

4) Friability. Different solids have different degrees of what is known as "friability," that is, their susceptibility to breaking up into smaller particles when subjected to impact and shear force. A high gas velocity and many bends in the transport system will tend to increase the likelihood that there will be a different particle size distribution as well as smaller particles in the receiving vessel.

5) Moisture Content. Moisture and humidity can affect solid particles' adhesion to each other. The humidity of the conveying airstream may need to be controlled and monitored. If a vacuum system is used, leakage of ambient air into the system may also bring moisture into the system.

This transport can be done in two different fashions, *dilute phase* and *dense phase*. Dilute phase refers to a condition where the solid being transported is suspended in the air. This may be preferable for lower-density solids being transported over long distances. When the gas velocity is high and the particle size small, a dust collection system is usually employed to prevent discharge into the atmosphere.

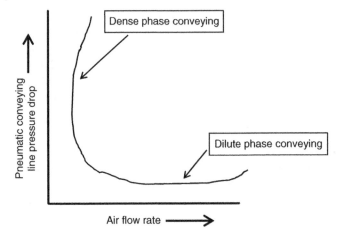

Figure 15.9 Pressure- versus vacuum-driven conveying systems.

A dense-phase system uses a small amount of gas and "pulses" the solids through the pipeline, with the pipeline nearly full of solids. This is more appropriate for denser solids and where concerns about friability and fines generation are present. There may still be a need for dust collection, but not nearly as large or sophisticated. There will also be less particle degradation, but more pressure drop.

Figure 15.9 shows the general relationship and extremes of the two types of design.

We can see the trade-off between airflow rate and pressure drop. Another important point is to not design or attempt to operate a pneumatic conveying system between these two extremes. This will cause pulsation and surging of flow, as the solids will sometimes be in the gas phase, and at other times will not.

Just as complicated piping networks need proper labeling and safety protocols to prevent liquids and gases from going into the wrong vessels, the same is true about pneumatic conveying systems. It is important to properly designate the routing of solids transport and provide safety interlocks so that the wrong product does not wind up in the wrong storage system or hopper car.

The energy used in a pneumatic conveying system will depend upon the amount of material being transported as well as upon whether the material is being elevated (such as into a storage silo from ground level).

Figures 15.10 and 15.11 show the energy used in such a system as a function of amount of material, length, and elevation.

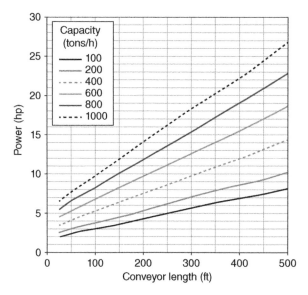

Figure 15.10 Pneumatic conveyor energy use versus length used. Source: Reproduced with permission of Engineering Toolbox.

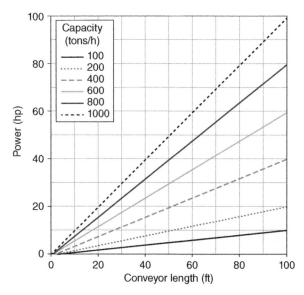

Figure 15.11 Pneumatic conveyor energy use versus lift used. Source: Reproduced with permission of Engineering Toolbox.

Solids Size Reduction Equipment

Frequently, especially in the mining of minerals and ores, the size of the solids is much greater than that desired in a downstream use. This requires that the raw solids be reduced in particle size, primarily through the use of mechanical energy.

There are many types of particle size reduction equipment, including hammer mills, rod mills, pulverizers, cage mills, roll crushers, attrition mills, and others. They each have their area of unique use and application with a great deal of overlap, normally requiring multiple vendor trials and evaluations. One common factor is the noise associated with these types of equipment, and it is normal to have advanced hearing protection and/or equipment isolation. An additional common factor is a relatively large energy use that increases as the desired particle size decreases.

We will review some of these in more detail. A hammer mill is basically a rotating shaft with hammer arms attached, but free to rotate. Solids size reduction is a function of starting particle size, hardness difference between the solid and the rotating arm, and the amount of time. A wood chipper would be a good analogy. A typical hammer, with its internals, is shown in Figure 15.12.

Another type of mill used to crush rocks and ores is a rod mill, shown in Figure 15.13.

Industrially, inside these mills are thousands of rods that are free to rotate and provide the impact force.

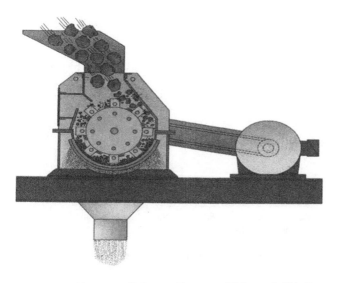

Figure 15.12 Hammer mill. Source: Courtesy of Schutte-Buffalo Hammermill.

Figure 15.13 Rod mill. Source: https://commons.wikimedia.org/wiki/File:Ball_mill.gif.

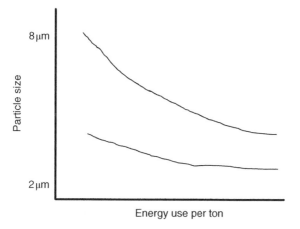

Figure 15.14 Relative energy input versus particle size required.

Attrition mills are similar in the fact that they use grinding media, but the media are typically spherical in shape and are used for pigments, graphite, food products, glass frits, rubber products, and cellulosic products. Both of these types of mills use a combination of compression and attrition forces to reduce particle size.

No matter what type of grinding equipment is used, the grinding media choice will be affected by variables such as initial/final particle size ratio, hardness of media versus material, potential discoloration concerns, contamination, and cost for media replacement. Energy use in any of these types of equipment will follow a general curve as shown in Figure 15.14.

Energy use increases dramatically as smaller particle size is required.

There are some specific laws that define energy consumption depending upon the range of particle size reduction. The first of these is Kick's law for large particle size greater than 50 μm:

$$E = K_1 \ln\left(\frac{D_{pi}}{D_{pf}}\right)$$

where E is proportional to the energy requirement, K_1 is an empirical constant, D_{pi} is the starting particle size, and D_{pf} is the final desired particle size. The energy requirement is proportional to the log of the particle size reduction requirement. This equation is a restatement of the difficulty of achieving small particle size.

For finer particles in the 0.5–50 μm range, a slight different equation normally applies, known as Bond's law:

$$E = K_2\left(\frac{1}{\sqrt{D_f}} - \frac{1}{\sqrt{D_i}}\right)$$

For very fine particles less than 0.05 μm, the energy requirement is represented by

$$E = K_3\left(\frac{1}{D_f} - \frac{1}{D_i}\right)$$

General requirements and concerns about all size reduction equipment include the following:

1) Speed of operation of the equipment, in terms of not only energy but also noise requirements and safety concerns.
2) Spacing between mechanical components of the various machines as they affect throughput of the material and potential contamination of the product with metal from the machines.
3) Particle size distribution will also most likely change in addition to the average particle size. This will affect downstream processing and storage.
4) The input of energy in all such equipment will raise the temperature of the solids being processed. It is important to calculate an accurate energy balance, monitor temperatures, and take into account possible product decomposition.
5) In addition to chemical decomposition concerns, we must never forget the high potential for serious injuries around high-speed rotating equipment. Proper guards, lockouts, and procedures must be in place.

As with many chemical engineering unit operations we have discussed, there are overlapping choices in equipment. In Table 15.3, we see a summary of the various types of solids size reduction equipment and their capabilities and limitations.

Table 15.3 Guide to size reduction equipment selection.

Table 2. Use this guide for equipment selection.										
Product particle size, μm	5,000	1,000	500	150	50	10	2	<1	Maximum Hardness	Reduction Ratio*
Crushers									Hard	10:1
Cutting mills/slicers									Soft	50:1
Pin/cage mills									Soft	25:1
Hammer mills									Intermediate	>50:1
Roll presses									Hard	10:1
Jet mills									Hard	>50:1
Media mills (tumbling and stirred)									Hard	>50:1

Shaded region indicates suitability of this type of mill for sizes larger than that shown.
* These are approximate maximum values. Size-reduction ratio will depend greatly upon the type of material.

Source: Chemical Engineering Progress, 4/16, pp. 48–55. Reproduced with permission of American Institute of Chemical Engineers.

As we can see, the hardness, the initial particle size, and the relative size reduction required are all important variables in choosing the optimum equipment. A number of vendor trials are normally necessary.

Cyclones

Frequently, in handling solids and transferring them via the various processes discussed, there will be a reduction in particle size in that process, and if a gas stream is used, it is often not allowed nor economical to discharge this stream into the atmosphere. Some type of dust collection system may be required.

The most common is a cyclone, no different in principle to the newer cyclonic vacuum machines used for home cleaning today. The gas stream is impinged against the inner wall of a cyclonic device, and the solids collected are sorted at the bottom discharge. A typical example is shown in Figure 15.15.

Many times, the inside of the cyclone contains a bag to prevent particle discharge, similar to a home vacuum cleaner.

From a practical standpoint, this kind of equipment is ideal in that there are no moving parts, except for a possible discharge or sealing device at the bottom. However, they have their limits (as do all types of process equipment). Their collection efficiency is based on the logarithmic difference in particle size they are able to collect, as shown in Figure 15.16.

To avoid unwanted discharge of solids into the atmosphere or downstream of the scrubbing device, it is critical to know where this "breakpoint" is on performance and operate well to the right of it as seen in Figure 15.16.

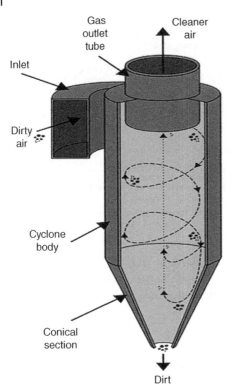

Figure 15.15 Dust cyclone. Source: By end:User:Cburnett [GFDL (http://www.gnu.org/copyleft/fdl.html) or CC-BY-SA-3.0 (http://creativecommons.org/licenses/by-sa/3.0/)], via Wikimedia Commons.

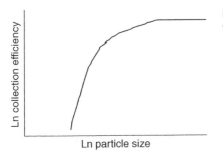

Figure 15.16 Cyclone collection efficiency versus particle size.

Screening

Once dried solids have been produced, it still may be necessary to screen them to isolate certain fractions by particle size and particle size distribution. This unit operation is frequently referred to as sieving, especially if its function is to recover particular solids sizes for use or sale. Solids size fractions are frequently expressed as "mesh size." Table 15.4 lists some of the conversions.

Table 15.4 Mesh and size conversion and examples.

US Standard	Space between wires		
Sieve mesh no.	Inches	Microns	Typical material
14	0.056	1400	
28	0.028	700	Beach sand
60	0.0098	250	Fine sand
100	0.0059	150	
200	0.0030	74	Portland cement
325	0.0017	44	Silt
400	0.0015	37	Plant pollen

Screeners use vibratory or centrifugal force to separate materials. The rate of solids handling and separation will be affected by:

1) Particle size and particle size distribution
2) Difference between screen hole size and the solids size/distribution
3) Vibratory speed
4) "Friability" of the solids (tendency to break up when subjected to force)

They can also be either batch or continuous operations.

Since vibratory screens involve the use of mechanical energy, their components will be subject to wear, and the screen elements will need to be inspected regularly to ensure there is no degradation in the screen hole size.

Hoppers and Bins

Once a solid material has been transported, separated, and classified, it is normally stored in a bin or silo prior to being loaded into hopper cars, bulk sacks, or other solids shipping containers. At the other end, the receiving customer must also be able unload and transport the solid material into their usable storage.

There are several key design and operational variables in the design of hoppers and bins for solids:

1) Solids Density. As opposed to liquids and gases, which have fixed densities if pressure and temperature are known, we characterize solids density in two ways. *Bulk density* would refer to the density (i.e., $\#/ft^3$ or ρ_B) of the solids upon loading into a vessel. *Tapped density* (ρ_T) would refer to the density after the material has been "vibrated" or "shaken." This density will

always be greater, and we see this phenomenon in our daily lives when we open a cereal box. The box is rarely full. It was most likely full when it was filled at the factory, but the contents settle during handling and transport. The ratio of the tapped density to the bulk density is called the *Hausner ratio* and would be $\rho T/\rho B$.

2) Shear Strength. This is a measure of the solid's ability to "stick to itself" or an inverse measure of how difficult it is to "shear" the solids into smaller particles. The reason this is important in the solids area is that, in the process of flowing and moving, solids are subjected to forces and these can cause solids to degrade and decrease in size. A material's resistance to shear would be expressed as force per unit area or #/ft^2, in a direction parallel to the solid's surface. The higher this number, the less likely it is that a solid will degrade into smaller particles when being transported or moved.

3) Tensile Strength. This refers to a solid's resistance to force perpendicular to its surface (think about tearing a sheet of paper). If this number is low, the solids will have more of a tendency to break apart when subjected to mechanical impact. This is an important parameter when considering how much a particle will degrade as it passes through transport devices.

4) Jenike Shear Strength. After solids are transferred to a hopper by a conveying system, their ability to be removed from the hopper or silo depends not only upon the previous variables but also upon the shear strength between the solids and the hopper or silo wall material. If a solid has a strong adhesion to the wall material, there will be a tendency for the solids to "hang up" along the wall and produce nonuniform flow out of the hopper, with solids in the middle of the hopper flowing out the discharge faster than those clinging to the walls.

One of the simple physical tests frequently done on a solid is to measure its "angle of repose," as illustrated in Figure 15.17.

Figure 15.17 Angle of repose of a solid material. Source: Captain Sprite, https://commons.wikimedia.org/wiki/File:Angleofrepose.png. Used under CC BY-SA 2.5 https://creativecommons.org/licenses/by-sa/2.5/deed.en. © Wikipedia.

The stronger the adhesion forces between the solid particles, the steeper this angle will be. We can observe this at home by simply observing the angle formed from a batch of sugar versus a batch of flour dumped on a kitchen surface. It is important to take into account the atmospheric conditions in measuring these variables. Humidity can affect many of these variables.

If these variables are not all taken into account, we can see any or all of the following results in an industrial setting:

1) No Flow. If the wall adhesion and cohesive strength of the solids is sufficient, these forces may be greater than the gravitational force trying to flow the materials downward.
2) Erratic Flow. If these balancing forces are close, there may be conditions where the solids flow and stop erratically.
3) Nonuniform Flow. The balance of these forces may be such that solids do flow, but due to differences in particle size, the uniformity of what leaves the hopper may be quite different than what has entered the hopper.
4) Flooding. In an exaggerated condition of erratic flow, the solids, after having hung up in the hopper due to adhesive forces, may suddenly surge from the hopper when the force balance is suddenly overcome.

The angle of the metal at the hopper discharge will be an indirect indication of the angle of repose. The steeper this angle is, the more difficult it is to obtain flow of solids from the hopper. It is also likely that if a hopper has been improperly designed, we would see a sledge hammer at the bottom of the hopper to be used by operating personnel to "release" the solids hung up in the hopper.

Solids Mixing

It is sometimes desired to blend and mix solids entering a silo or hopper storage system or to mix solids being discharged from several hoppers. In these cases, it is critical that the properties of the actual solids mixture be used to design the transfer equipment and not to mathematically average what we think the individual properties are. This kind of mixing can be done with multiple-screw conveyor systems or a vertical cone mixing system.

Some of the key points are summarized as follows:

1) Solids are unique from the standpoint of chemical engineering in that their properties and processability and handling depend on more than their chemical formula. Variables such as particle size and particle size distribution are key variables that are not of concern when processing typical gases and liquids.
2) These unique solids properties are greatly affected by the equipment used to process, transfer, and store them. Mistakes are often made in designing or

specifying equipment based on the input to a solids handling device as opposed to the output from it.

3) There is little or no college education in this area, resulting in the majority of working and theoretical knowledge being in the hands of vendors and specialized consultants, both within companies and outside.

4) The flammability and fire hazards of solids are frequently not understood or ignored. Dust explosions can cause just as much damage and loss of human life as a petrochemical fire and explosion. Many types of solids handling equipment will input a great deal of energy into a solids system, requiring a thorough understanding of the heat balance associated with this heat input.

5) There tend to be a large number of equipment options for various types of solids processing and handling, requiring a thorough understanding of process and product requirements and the ability to make choices in an unbiased manner.

Solids and Coffee Brewing

Coffee grounds and beans are solids, so everything reviewed here is applicable, though not necessarily visible to the consumer. Beans are harvested from trees and need to be transported and stored. Then they are shipped to a coffee roasting (chemical reactor) operation. This may be the end of the journey as they are packaged into vacuum bags (to slow down the chemical flavor degradation) and shipped to a warehouse and ultimately to the grocery store shelf.

The beans are more usually ground and packaged into vacuum bags or cans (same reason) and then the same journey to the warehouse and grocery store shelf. The handling of large quantities of ground coffee will involve virtually all of the solids transport and storage unit operations we have discussed. The degree to which the particle size is reduced will determine whether the coffee grounds are classified as "perc," "drip," or "espresso." The finer the particle, the more surface area available for the water to contact in the brewing process and the stronger the taste of the coffee brewed. Many fussy coffee brewers will grind the beans at the last minute to minimize the surface area available for oxidation degradation (kinetics and reaction engineering).

Discussion Questions

1 To what degree are solids handling important in your processes? Raw materials? Intermediates? Final products?

2 Are all the appropriate information on the various solids known? How do they affect the design of solids handling equipment?

3 Are any of your customers complaining about caked products in drums or hopper cars that cannot be easily emptied? Why is this a problem? Is this a new phenomenon? If so, what changed?

4 If your customer were to request a different dryness of product, would you know how to produce it? The cost of doing so? What new types of equipment might be evaluated?

Review Questions (Answers in Appendix with Explanations)

1 The energy used in particle size reduction is primarily a function of:
  A __Price of energy
  B __Ratio of incoming particle size to exiting particle size
  C __Size of the hammers or pulverizers
  D __Strength of the operator running the equipment

2 Cyclones have primarily one very positive design feature and one very negative design feature:
  A __No moving parts and sharp cutoff in particle separation
  B __No motor and low particle collection
  C __Can be made in the farm belt but cannot collect large size corn cobs
  D __Are small and can make a lot of noise

3 A key solids characteristic in assessing solids cohesion is:
  A __Particle size
  B __Height of solids pile
  C __Slope of laziness
  D __Angle of repose

4 A poorly designed hopper can cause:
  A __No flow when the bottom valve is opened
  B __Segregated particle size flow
  C __Surges in flow behavior
  D __All of the above

Additional Resources

Note: Due to the unique nature of solids handling equipment, several videos are listed at the end of this resources list to allow visualization of the working of some of the equipment discussed. Some of these videos are commercially produced. Neither the author, AIChE, nor Wiley endorses any of the particular equipment demonstrated.

Alamzad, H. (2001) "Prevent Premature Screen Breakage in Circular Vibratory Separators" *Chemical Engineering Progress*, 5, pp. 78–79.

Armstrong, B.; Brockbank, K. and Clayton, J. (2014) "Understanding the Effects of Moisture on Solids Behavior" *Chemical Engineering Progress*, 10, pp. 25–30.

Carson, J.; Troxel, T. and Bengston, K. E. (2008) "Successfully Scale Up Solids Handling" *Chemical Engineering Progress*, 4, pp. 33–40.

Maynard, E. (2012) "Avoid Bulk Solids Segregation Problems" *Chemical Engineering Progress*, 4, pp. 35–39.

Mehos, G. (2016) "Prevent Caking of Bulk Solids" *Chemical Engineering Progress*, 4, pp. 48–55.

Mehos, G. and Maynard, E. (2009) "Handle Bulk Solids Safely and Effectively" *Chemical Engineering Progress*, 09, pp. 38–42.

Zalosh, R.; Grossel, S.; Kahn, R. and Sliva, D. (2005) "Safely Handle Powdered Solids" *Chemical Engineering Progress*, 12, pp. 22–30.

Videos of Solids Handling Equipment

https://www.youtube.com/watch?v=WFE-vPXxxXc (accessed August 27, 2016).
https://www.youtube.com/watch?v=g7DdLLPknDo (accessed August 27, 2016).
https://www.youtube.com/watch?v=TdIq4WR50jQ (accessed August 27, 2016).
https://www.youtube.com/watch?v=L6sgGXXYdEU (accessed August 27, 2016).

16

Tanks, Vessels, and Special Reaction Systems

The actual physical design and detailed specification of tanks and vessels (wall thickness, exact metal alloy, etc.) are done by mechanical and structural engineers. However, chemical engineering input is required to decide the general geometry and geometrical ratios, pressure and vacuum requirements, and materials needed to resist corrosion of the liquids, solids, and gases being contained.

Categories

Tanks and vessels can be divided into three broad categories:

1) Storage or Batch Vessels. In this case we are using a tank or vessel to store a raw material, an intermediate, or a final product. Storage can refer to raw materials, intermediates, final products, and inventory. It can also refer to "day tanks" where a "lot" of material is quarantined for analysis prior to being forwarded for further processing. This type of tank can also be used as an intermediate process point for settling layers or letting solids or "rag layers" to settle out. Occasionally, these tanks may be used as precipitators after a reaction or crystallization. Though these situations and uses may appear to be benign, safety issues that must be considered include leaks, overflowing, product contamination, and corrosion.

2) Pressure and Vacuum. Most storage tanks operate under atmospheric pressure unless the contents are a liquefied compressed gas. Because of the typical conservative design of such tanks, these tanks can normally handle a small amount of pressure. However, these types of tanks have little or no capability to handle vacuum and can collapse under vacuum. If the tank is expected to store materials under pressure, its pressure rating must accommodate this. It is essential that vents from a tank, if designed to maintain atmospheric pressure, do not become closed or clogged. This can happen

Chemical Engineering for Non-Chemical Engineers, First Edition. Jack Hipple.
© 2017 American Institute of Chemical Engineers, Inc. Published 2017 by John Wiley & Sons, Inc.

because covered vents are not uncovered after maintenance or, as shown in this illustration, a vent is plugged by an insect nest! (Figure 16.2).
3) Process and Agitated Tanks. This would refer to reaction vessels and storage vessels that are agitated. Most of the time we are concerned about process reaction vessels. We discussed reaction rates and kinetics earlier, and the design of an actual chemical reactor incorporates these fundamentals into its design. There is an overlap between these two basic concepts in some situations such as crystallization and settling.

Corrosion

In addition to the normal concerns about corrosion to the tank from its contents and the outside environment, tanks that are situated at ground level in contact with soil are subject to what is known as galvanic corrosion, where the tank metal, in contact with moisture in the soil, becomes part of an electrochemical circuit (a battery). It is a normal practice to monitor this and provide a reverse current to counteract this current to ensure that the bottom sections of the tank (frequently unseen) do not collapse or leak.

The physical design of any tank or vessel includes not only such obvious items such as volume and pressure rating (especially if used as a reactor, which handles or generates gases) but also such geometric factors such as diameter, height-to-width ratio, the nature of the flanges, wall thickness (including corrosion concerns), materials of construction, and the nature of jackets that might be used around the tank for cooling or heating. Mechanical and civil engineers will be involved in ensuring that a vessel meets codes and that welding done during tank construction and assembly meets appropriate codes.

It is tempting to think of tanks and storage vessels as nonhazardous equipment. However, when such tanks and vessels contain large volumes of hazardous materials, release of these materials can cause serious safety incidents. The Center for Chemical Process Safety (http://www.aiche.org/ccps), a consortium of over 150 chemical companies, which studies and provides learning examples for chemical engineers, reported this major incident in England in 2005, which provides a classic example of the hazards of simple storage tanks (Figure 16.1).

The learnings from the incident are discussed in detail in the Summary section, but let's review some of the following key points:

1) It is critical to know the inventory of a tank at all times for a number of important reasons:
   a) Overfilling or under filling of a tank, resulting in overflow or damage to pumps associated with the storage tank. Overfilling can also result in discharge of flammable materials (which will ultimately find an ignition source),

Messages for Manufacturing Personnel

http://www.aiche.org/ccps/safetybeacon.htm

September 2006

Overfilling Tanks — What Happened?

Photograph courtesy of Royal Chiltern Air Support Unit.

On Sunday, December 11, 2005, gasoline (petrol) was being pumped into a storage tank at the Buncefield Oil Storage Depot in Hertfordshire, U.K. At about 1:30 A.M., a stock check of the tanks showed nothing abnormal. From about 3 A.M., the level gage in one of the tanks recorded no change in reading, even though flow was continuing at a rate of about 550 m³/h (2,400 U.S. gal/min). Calculations show that the tank would have been full at about 5:20 A.M., and that it would then overflow. Pumping continued, and the excess gasoline overflowed from the top of the tank and cascaded down the sides, forming a liquid pool and a cloud of flammable gasoline vapor. At about 6 A.M., the cloud ignited, and the first explosion occurred. This was followed by additional explosions and a fire that engulfed 20 storage tanks. Fortunately, there were no fatalities, but 43 people were injured. Approximately 2,000 people were evacuated. There was significant damage to property in the area, and a major highway was closed. The fires burned for several days, destroying most of the site and releasing large clouds of black smoke, which impacted the environment over a large area.

Did You Know?

Photo courtesy of Royal Chiltern Air Support Unit.

▶ Overfilling of process vessels has been one of the causes of a number of serious incidents in the oil and chemical industries in recent years — for example, the explosion at an oil refinery in Texas City, TX, in March 2005.

▶ The tank involved in this incident had an independent high-level alarm and interlock, but these components did not work. The cause of the failure is still under investigation.

▶ A spill of flammable material, such as gasoline, can form a dense, flammable vapor cloud. This cloud can grow and spread at ground level until it finds an ignition source. This ignition source can cause the cloud to explode.

What Can You Do?

Photo courtesy of Hertfordshire Constabulary.

▶ When you transfer material, make sure that you know where the material is going.

▶ When you are pumping into a tank, and the level or weight indicator in that tank does not increase as you would expect, stop the transfer and find out what is happening.

▶ Make sure that all safety alarms and interlocks are tested at the frequency recommended in the plant process-safety-management procedures.

▶ If you have alarms and interlocks that are not regularly tested, ask the plant process safety manager if they are safety critical and whether they should be on a regular testing program.

Read the reports about this incident at:
http://www.buncefieldinvestigation.gov.uk

If you are pumping material, be sure you know where it is going!

The Beacon is usually available in Arabic, Chinese, Dutch, English, French, German, Gujarati, Hebrew, Hindi, Italian, Japanese, Korean, Portuguese, Spanish, Swedish and Thai. Circle No. 103 on p. 57 for a free electronic subscription to the Beacon.

CEP   September 2006   www.aiche.org/CEP   **17**

Figure 16.1 Consequences of overfilling a tank. Source: Chemical Engineering Progress 7/06, p17. Reproduced with permission of American Institute of Chemical Engineers.

a release of material that will cause environmental damage, or human exposure. There are many state and federal regulations regarding diking requirements, but not having any leaks at all should be the primary objective.

b) If the tank contents and volume are not known accurately, upstream and downstream processing units can be negatively affected.

c) The pressure within the tank or vessel must also be known for two primary reasons. The first, the one that immediately comes to mind, is rupture of a vessel by exposing it to an internal pressure greater than its design. However, the second and just as important a consideration is the potential exposure of tank contents to vacuum, as mentioned earlier. Atmospheric pressure is NOT zero—it is approximately 14.7 psig at sea level and slightly lesser at high altitudes. Many tanks and vessels can withstand a moderate amount of pressure but cannot withstand even a small amount of *external* pressure (i.e., the difference between atmospheric pressure and vacuum in the tank). Figure 16.2 (from the Center for Chemical Process Safety publication, the Beacon) shows the consequences of plugging a vent from a vessel not designed to handle vacuum. This can happen because covered vents are not uncovered after maintenance or, as shown in Figure 16.2, a vent is plugged by an insect nest!

Agitated tanks and vessels are used for many purposes, including reactions, crystallizers, blend tanks, and precipitators. A general diagram of an agitated vessel is shown in Figure 16.3.

From a physical design standpoint, there are a number of parameters that must be chosen:

1) Diameter and Height. The product of these will determine the overall volume of the tank. It is important, when designing a reactor tank's volume, to take into account the gas volume, which may be from a reactant or an evolved gas from a reaction. If a batch reaction is the primary function of the vessel, the volume needed must also be taken into account as well as the time spent filling, draining, and cleaning the reactor vessel.

2) Height-to-diameter Ratio (Z/T). This refers to the ratio of the height of liquid in the tank to the inside tank diameter. This ratio will affect the physical stress on the agitator system and its motor and drive system. For example, if the vessel is very short and wide in diameter, there will be a great deal of stress on the shaft in the latitudinal direction and the agitation system may have difficulty in mixing side to side. At the other extreme, if the vessel is very tall and skinny, there will be excellent sideways mixing, but poor top-to-bottom mixing, and the longitudinal stress on the agitator shaft will be large. These effects will be exaggerated if baffles are attached to the shaft or if the liquid viscosity and densities are high, requiring more energy to mix the fluids.

Messages for Manufacturing Personnel
http://www.aiche.org/CCPS/Publications/Beacon/index.aspx

Vacuum Hazards — Collapsed Tanks

February 2007

The tank on the left collapsed because material was pumped out after somebody had covered the tank vent to atmosphere with a sheet of plastic. Who would ever think that a thin sheet of plastic would be stronger than a large storage tank? But, large storage tanks are designed to withstand only a small amount of internal pressure, not vacuum (external pressure on the tank wall). It is possible to collapse a large tank with a small amount of vacuum, and there are many reports of tanks being collapsed by something as simple as pumping material out while the tank vent is closed or rapid cooling of the tank vapor space from a thunder storm with a closed or blocked tank vent. The tank in the photograph on the right collapsed because the tank vent was plugged with wax. The middle photograph shows a tank vent that was blocked by a nest of bees! The February 2002 Beacon shows more examples of vessels collapsed by vacuum.

Did you know?

▶ Engineers calculated that the total force from atmospheric pressure on each panel of the storage tank in the left photograph was about 60,000 lbs.
▶ The same calculation revealed that the total force on the plastic sheet covering the small tank vent was only about 165 lbs. Obviously this force was not enough to break the plastic, and the tank collapsed.
▶ Many containers can withstand much more internal pressure than external pressure — for example, a soda can is quite strong with respect to internal pressure, but it is very easy to crush an empty can.

What can you do?

▶ Recognize that vents can be easily blocked by well intended people. They often put plastic bags over tank vents or other openings during maintenance or shutdowns to keep rain out of the tank, or to prevent debris from entering the tank. If you do this, make sure that you keep a list of all such covers and remove them before startup.
▶ Never cover or block the atmospheric vent of an operating tank.
▶ Inspect tank vents routinely for plugging when in fouling service.

Vacuum — it is stronger than you think!

The Beacon is usually available in Arabic, Chinese, Dutch, English, French, German, Gujarati, Hebrew, Hindi, Italian, Japanese, Korean, Portuguese, Spanish, Swedish, and Thai. For more information, circle 103 on p. 63.

Figure 16.2 Consequences of pulling vacuum on storage vessel. Source: Chemical Engineering Progress, 3/10, pp. 25–32. Reproduced with permission of American Institute of Chemical Engineers.

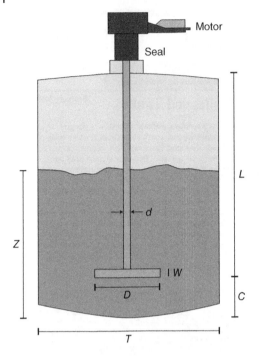

Figure 16.3 Agitated vessel parameters. Source: Chemical Engineering Progress, 8/15, pp. 35–42. Reproduced with permission of American Institute of Chemical Engineers.

3) Impellers. Figure 16.3 shows only one impeller near the bottom of the agitator shaft. There could be multiple impellers installed vertically. The number needed will be a function of the physical property differences, the Z/T ratio in the tank, and the agitator horsepower available.

As a rule of thumb, 20% of the volume in a tank is left for gas volume and the diameter-to-height ratio is between 0.8 and 1.4. Whatever choices are made in the agitator system, they must meet the standard horsepower and gear box requirements.

If a tank contains solids, the settling rate of these materials may be important and may indirectly impact any agitation design. The settling rate will not only be a function primarily of density difference and solids concentration, as shown in Figure 16.4, but will also be affected by viscosity.

There are numerous types of agitators used in vessels including the following:

1) Marine Propellers. These are commonly used in boats and are fairly expensive since they are made from a casting process (Figure 16.5).
2) Rushton Turbine Agitators. These were developed as an alternative to cast mixers and propellers and are multipurpose and multifunctional agitators whose geometry can be varied and adjusted to suit various process conditions. They are also relatively inexpensive to manufacture. Example of such agitators are shown in Figure 16.6.

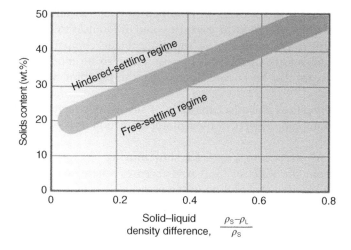

Figure 16.4 Settling versus density and solids concentration. Source: Chemical Engineering Progress, 1/14, pp. 30–36. Reproduced with permission of American Institute of Chemical Engineers.

Figure 16.5 Propeller agitator. Source: PublicDomainPictures.com.

Figure 16.6 Rushton turbine agitators. Source: Reproduced with permission of Yahoo.

The pitch and number of blades can be modified to adapt to different mixing requirements, different liquid-to-gas ratios, and differences in liquid viscosities and densities. The diameter of the plane blade compared to the shaft diameter and nature of the attachments can also be modified. The attached blades can also be curved, increasing the pumping versus agitation ratio.

These geometric differences will affect the ratio of side-to-side (agitation) and top-to-bottom (pumping) aspects of the agitator's function as shown in Figure 16.7.

It is also possible to insert into a vessel a slow moving "scraper" that will remove solids from the walls of the vessel in cases where solids are precipitating out of a reaction and coating of the vessel walls would inhibit heat transfer.

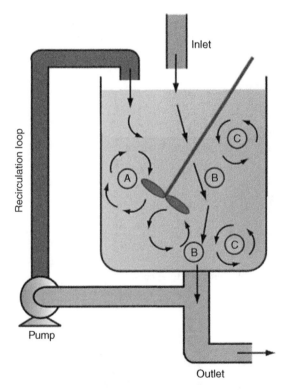

Figure 16.7 Flow patterns induced by agitation: side to side and top to bottom. Source: Chemical Engineering Progress, 3/10, pp. 25–32. Reproduced with permission of American Institute of Chemical Engineers.

Heating and Cooling

When reactor vessels and tanks must be heated or cooled, there are three basic options as follows:

1) Jacket of the Vessel. Steam or a cooling fluid can be injected into the jacket of the vessel. The rate of heat transfer will be governed by the general heat transfer equation we have discussed previously:

$$Q = UA\Delta T$$

   In the case of a stirred tank, the overall heat transfer coefficient, U, will be affected by the properties of the fluid (viscosity, density, heat capacity) as well as the agitation rate. If a chemical reaction is being carried out in the vessel, these properties will change with time as the chemical reaction proceeds (with physical properties changing), and these differences must be taken into account. If gases are evolved during the reaction, this will make the heat transfer calculations more complex.

2) Coils Inserted into the Vessel. The same issues apply as mentioned in the case of the use of jackets.

3) Direct Steam or Cooling Injection. As long as the heat and material balance issues are taken into account, it is possible to directly inject steam (for heating) or water into a reactor for heating or cooling. Dilution of the process stream must be taken into account. Since there is no material barrier between the steam and the material being heated, the heat transfer rates are much higher than indirect heating mechanisms.

4) Addition of turbulence into the jacket of the reactor and internal baffling can also be considered as seen in Figure 16.8.

Power Requirements

The amount of power required for a tank or a vessel agitator will be affected by many factors including the following:

1) Amount of agitation required
2) Physical properties of the liquids or gases in the tank or vessel (density, viscosity)
3) Ratio of gas to liquid and gas/liquid to solids (if any)
4) Geometric design of the tank

There are several different equations used to calculate the relationships between physical properties and turbulence, agitation, and energy requirements

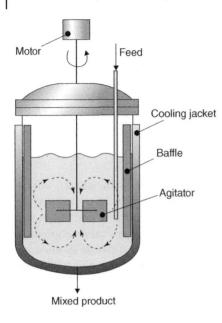

Figure 16.8 Agitated, baffled, and jacketed reactor vessel. Source: Reproduced with permission of Sulzer.

in an agitated tank system. The turbulence in the tank (the impeller Reynolds number) can be estimated by

$$N_{Re} = \frac{D^2 N \rho}{\mu}$$

where D is the impeller diameter, N is the shaft speed, ρ is the fluid density, and μ is the liquid viscosity. As we increase the diameter, rotational speed, and density, the turbulence increases; as the viscosity increases, it decreases. It is important to calculate these numbers based on the properties and conditions in the vessel at the time of concern and not the feed materials.

We will also see a number called the impeller "pumping number":

$$N_p = \frac{P}{\rho N^3 D^5}$$

This equation shows the strong dependency of power requirement on shaft speed and impeller diameter. This relationship explains, in part, the reason why extreme vessel and tank designs are not used (tall and skinny or short with wide diameter).

Tank agitation can also be done with a combination of mechanical agitation and aeration (or other gas) flow. The power used by the agitator plus the power used by the gas compression system may result in a minimum energy cost point as shown in Figure 16.9.

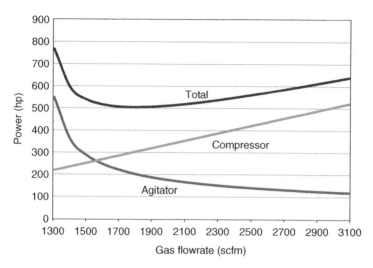

Figure 16.9 Minimal power consumption. Source: Chemical Engineering Progress, 8/15, pp. 35–42. Reproduced with permission of American Institute of Chemical Engineers.

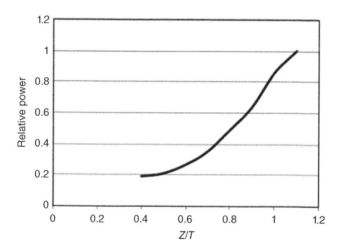

Figure 16.10 Power requirements versus liquid level. Source: Chemical Engineering Progress, 8/15, pp. 35–42. Reproduced with permission of American Institute of Chemical Engineers.

Agitation power requirements are also affected by the liquid level in the tank as shown in Figure 16.10.

Power requirements are also affected by the number of baffles and the Z/T ratio in the vessel, as summarized in Table 16.1.

Table 16.1 Relative power requirements versus Z/T ratio and number of impellers.

		Table 3. Looking at relative torque as a function of Z/T can also help you determine power requirements.					
		Square batch power split basis					
	Number of impellers	50/50		70/30		90/10	
Z/T		Relative torque	Relative power	Relative torque	Relative power	Relative torque	Relative power
0.5	1	1.00	0.79	1.41	1.12	1.80	1.43
1	2	1.00	1.00	1.00	1.00	1.00	1.00
1.5	3	1.00	1.15	0.86	0.99	0.73	0.84
2	4	1.00	1.26	0.80	1.00	0.60	0.75
2.5	5	1.00	1.36	0.75	1.02	0.52	0.70

Source: Chemical Engineering Progress, 2/12, pp. 45–48. Reproduced with permission of American Institute of Chemical Engineers.

Using tanks as blending systems is frequently done, and the time necessary to blend two different fluids will be affected by their differences in viscosity and density, as well as the mechanical design of the agitator and input power.

Tanks can also be used as settling vessels to allow solids to settle out after reactions or prior to filtration. As the solids concentration increases, the settling rate decreases as the solids settling will hinder each other. This phenomenon is especially visible at solids concentrations above 40% and with high viscosity fluids.

Flow patterns with a mixing vessel are important to consider, especially if uniformity is critical. The mixing within a vessel from top to bottom will require higher energy input than simple side-to-side mixing. If a gas is being generated or used as a reactant, or if a gas is being used as part of the agitation system, there will be an optimum ratio of agitator power to gas velocity, which will minimize overall power requirement.

Tanks and Vessels as Reactors

When tanks are used as reaction vessels, sizing is important and also the total cycle time. When a vessel is being used as a batch reactor, the size required must take into account not only the reaction time but also the fill time, drain time, and cleaning time that may be required. Occasionally we will see a tank reactor used in what would be described as a "semi-continuous" mode. This refers to the input into the reactor of reacting components and the reaction occurs during the filling process. This would normally be a fast reacting system. When the vessel is full, the flows are stopped and the products removed.

Another common use of tanks and vessels in reaction systems is what is known as a continuously stirred tank reactor (CSTR). In this case, reactants flow into the vessel continuously, reactants have a certain residence time in the vessel for the reaction to occur, and the flow constantly leaves the reactor. If we take a look at the previous illustration of a tank design, we can focus on how the feed is introduced. Depending on physical property differences (viscosity, density) and if any of the feed materials are gases as opposed to liquids, the feed could be introduced as shown in Figure 16.11 (near the impeller) to maximize immediate mixing.

A cooling jacket could also be used on the reaction vessel if the reaction is exothermic and needs cooling, or if the agitation energy needs to be balanced). Baffles can be used to improve mixing within the vessel. The required size of the reaction vessel will be related to the kinetics of the reaction(s) that are occurring. The slower the reaction rate, the larger the vessel volume will need to be. It is also possible to configure cooling or heating of such a system via external cooling or internal heat exchange piping.

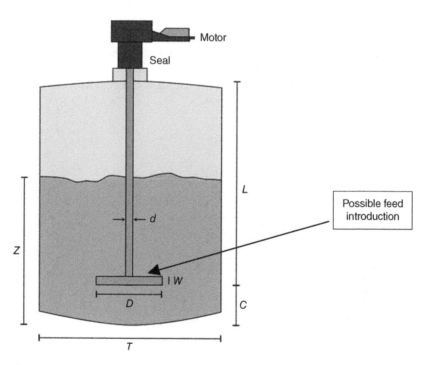

Figure 16.11 Feed introduction into an agitated vessel. Source: Reproduced with permission of Sulzer.

Static Mixers

If the reaction rates are extremely fast or the physical property differences between materials are minimal, it is possible for a static mixer to be used to mix and/or react materials within a pipeline, as shown in Figure 16.12.

These in-line mixers provide very intense mixing in a short piping length with the primary drawback being high pressure drop. These devices are suitable for in-line mixing of solutions and liquids of different compositions and physical properties as well as to conduct chemical reactions, precipitations, and dissolutions when the required time is very short. The primary downside to these devices is their very high pressure drop. A secondary concern may be erosion to the inserts if the liquids or slurries being processed have a high abrasive content.

Summary

Tanks and vessels have many uses within the chemical process industries, including storage, decanting, settling, product isolation for analysis, and chemical reaction systems. Their specific design will be affected by pressure and agitation requirements, gas/liquid/solids ratios, vessel codes, and the nature of the reaction kinetics if used as reactors. The detailed design of vessels and tanks must meet various codes specified by ASME. In-line mixers are low cost alternatives to tank mixing when lot isolation is not required and high pressure drop is not a barrier to their usage.

Figure 16.12 Static mixer. Source: Reproduced with permission of Sulzer.

Coffee Brewing Tanks and Vessels

The two most obvious are the carafe into which the brewed coffee drops and the cup that holds the final cup of coffee that is consumed. Carafes can be either glass or metal. Glass can easily break and protection against thermal burns is advisable, no different than if a similar situation existed in industry. Glass is usually corrosion resistant but can build up coatings that need to be removed. A metal container will be safer but will certainly have a corrosion rate. In the case of hot coffee, this is probably so small that it is of no practical concern, but it is not zero.

Previously ground coffee is occasionally stored in vacuum containers. These must be strong enough not to collapse under atmospheric pressure.

The final coffee cup is an agitated vessel if cream and sugar/sweetener (solids!) are added. The uniformity of the coffee flavor will be a function of how long the agitation lasts and what kind of agitator is used (spoon, stick, etc.).

Discussion Questions

1 How are tanks and agitated vessels used in your processes? How was the design and type of agitation chosen? Have these decisions been reviewed recently in light of new types of equipment availability?

2 What is the response of the agitation system to changes in physical properties (density, viscosity)?

3 Could static mixers replace tank mixers?

4 Is the possibility of vacuum collapse of storage vessels been reviewed?

5 Is there a more optimum combination of heat transfer and tank agitation that could be more optimal?

6 Have wall thicknesses of tanks and vessels been checked? How often?

Review Questions (Answers in Appendix with Explanations)

1 Simple storage tanks can be hazardous due to:
   A __Leakage
   B __Overfilling and under filling
   C __Contamination
   D __All of the above

2 A key design feature in an agitated vessel (its Z/T ratio) is:
  A __Its height-to-diameter ratio
  B __Which company manufactured it
  C __The engineer who designed it
  D __When it was placed into service

3 Choosing an agitator for a tank will be affected by:
  A __Densities of liquids and solids
  B __Viscosity of liquids
  C __Ratios of gases/liquids/solids
  D __All of the above

4 The significance of top-to-bottom agitation will be affected by:
  A __Gas and liquid densities and their changes over time
  B __Formation of solids as a reaction proceeds
  C __Necessity for liquid/solid/gas mixing
  D __Any of the above

5 Shaft horsepower requirements for an agitated vessel will be MOST affected by:
  A __Air flow available and ability to clean an exiting air stream
  B __Physical properties of materials being mixed and agitated
  C __Particle size degradation
  D __All of the above

Additional Resources

Amrouche, Y.; DavÈ, C.; Gursahani, K.; Lee, R. and Montemayor, L. (2002) "General Rules for Above Ground Storage Tank Design and Operation" *Chemical Engineering Progress*, 12, pp. 54–58.

Benz, G. (2012) "Cut Agitator Power Costs" *Chemical Engineering Progress*, 11, pp. 40–43.

Benz, G. (2012) "Determining Torque Split for Multiple Impellers in Slurry Mixing" *Chemical Engineering Progress*, 2, pp. 45–48.

Benz, G. (2014) "Designing Multistage Agitated Reactor" *Chemical Engineering Progress*, 1, pp. 30–36.

Dickey, D. (2015) "Tacking Difficult Mixing Problems" *Chemical Engineering Progress*, 8, pp. 35–42.

Garvin, J. (2005) "Evaluate Flow and Heat Transfer in Agitated Jackets" *Chemical Engineering Progress*, 8, pp. 39–41.

Machado, M. and Kresta, S. (2015) "When Mixing Matters: Choose Impellers Based on Process Requirements" *Chemical Engineering Progress*, 7, pp. 27–33.

Milne, D.; Glasser, D.; Hildebrandt, D. and Hausberger, B. (2006) "Reactor Selection: Plug Flow or Continuously Stirred Tank?" *Chemical Engineering Progress*, 4, pp. 34–37.

Myers, K.; Reeder, M. and Fasano, J. (2002) "Optimizing Mixing by Choosing the Proper Baffles" *Chemical Engineering Progress*, 2, pp. 42–47.

Post, T. (2010) "Understanding the Real World of Mixing" *Chemical Engineering Progress*, 3, pp. 25–32.http://en.wikipedia.org/wiki/Static_mixer (accessed August 26, 2016).

17

Chemical Engineering in Polymer Manufacture and Processing

Polymers have unique properties from a chemical engineering standpoint. They are long chain molecules with unique physical properties, and these properties change with the nature of the polymer being manufactured, how it is manufactured, and how it is used. Different polymers can be blended together to achieve certain properties. In addition, different monomers (the building blocks from which polymers are made) can be combined in many different ways to achieve further differentiation in product performance. When we formulate polymers with additives and pigments, we usually use the term plastics, but it is common to see both terms used to describe the starting polymer material.

What are Polymers?

Basically, polymers are products produced from joining together chains of reactive monomers, the starting molecules. One type of "joining" process is initiated with energy to activate a double bond in the starting molecule, allowing additional monomers to react with it, creating long chains of monomers. The length of this chain can be controlled and limited. A common term used is the "molecular weight" (MW) of the polymer.

In one example, the double bond within ethylene (CH_2=CH_2) is activated by heat or a catalyst to produce a monomer that we would describe as (CH_2–CH_2^*). This activation allows the ethylene to react with other ethylene molecules multiple times and produce "polyethylene," which we describe as (CH_2–CH_2–CH_2–CH_2)$_x$ where "x" is the number of ethylene monomers joined together. We commonly refer to "x" as the MW of the polymer. The higher the MW, the higher the "melting point" of the polymer will be and the higher will be its viscosity. Its mechanical strength, in general, will also be higher. There are numerous types of monomers that can be polymerized

Chemical Engineering for Non-Chemical Engineers, First Edition. Jack Hipple.
© 2017 American Institute of Chemical Engineers, Inc. Published 2017 by John Wiley & Sons, Inc.

including propylene, styrene, butadiene, and vinyl chloride, all having double bonds that can be activated. All the chains in a polymer system do not necessarily have the same chain length. There can be a variety and distribution of chain lengths. We use the term *molecular weight distribution* (MWD) to describe this. The more even this characteristic is, the more uniform the polymer properties will be under processing and use conditions. MWD can be viewed in the same way the particle size distribution was seen in the discussion on solids handling (see Figure 17.1).

Polymers, when made, can also display a property known as *crystallinity* meaning that portions of the long polymer chains have grouped together in such a way that there are structured domains of polymers within the overall polymer structure as shown in Figure 17.2.

When this type of characteristic is present, there will be multiple softening points as the polymer is heated and the various regions reach their softening points.

In addition to higher thermal stability than their starting monomers, polymers will also have some degree of thermal, chemical, and abrasion resistances. Their melting points will vary with the monomer used as well as the MW, and these two variables will affect their uses. Polymers are used in the manufacture of plastic garbage bags, blood tubing, fuel-resistant parts in automobiles, bake ware, kitchenware, and piping. In modified forms they are also used as foamed insulation, carpet backings, and house siding.

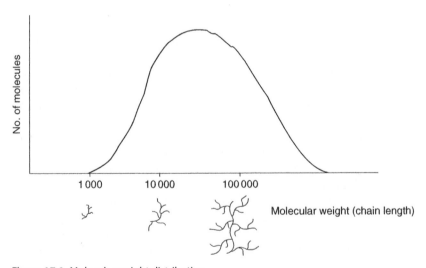

Figure 17.1 Molecular weight distribution.

Figure 17.2 Crystallinity areas in polymers.

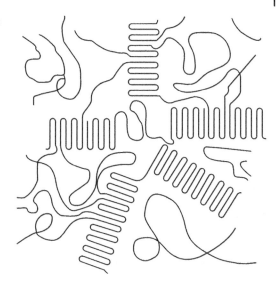

Polymer Types

In addition to a basic linear polymer, there are other possible types of polymers, described by a combination of their make up or how they are manufactured.

1) Linear versus Branched Polymers. The method of polymerization (primarily the catalyst used) can control whether the monomers are linked together in a linear fashion or in a branched fashion:

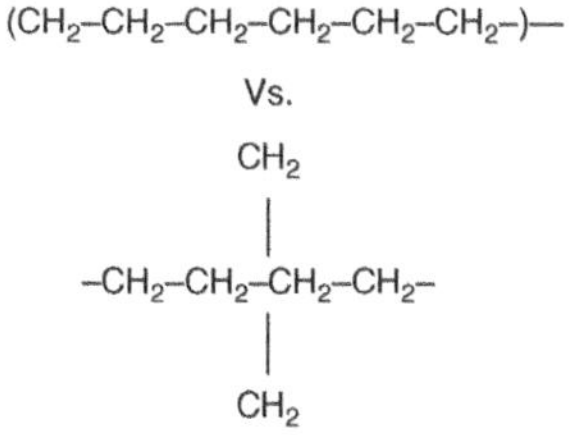

Figure 17.3 Linear versus branched polymers.

$$(CH_2-CH_2-CH_2-CH_2-CH_2-CH_2-)-$$

Vs.

$$CH_2$$
$$|$$
$$-CH_2-CH_2-CH_2-CH_2-$$
$$|$$
$$CH_2$$

The MW may be the same, but the branched structure produces a polymer of lower density and greater strength per unit volume. The change from linear to branched structure was a major breakthrough in polyethylene manufacturing and is the primary reason why typical trash bags, made from polyethylene, use far less material to achieve the same strength, compared to decades ago.

2) Copolymers. It is possible to copolymerize two different monomers, both of which having a reactive bond, as previously discussed, or a reactive chemistry

between the molecules. Combinations of styrene and butadiene (the basis for many synthetic rubber materials) are an example of this:

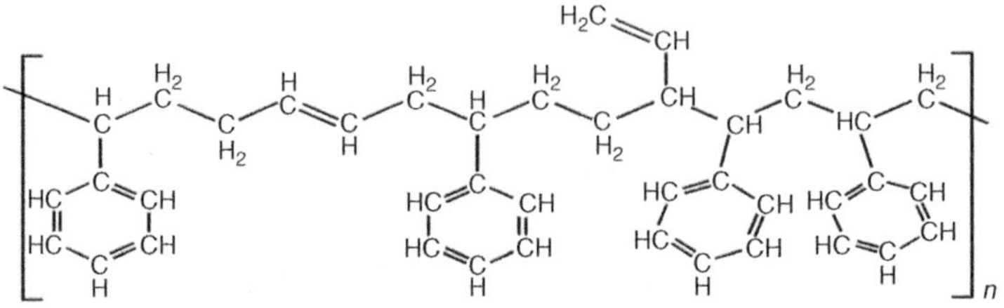

Figure 17.4 Styrene–butadiene copolymer.

In addition to the ratio of the two monomers, it is also possible to change the manner in which the monomers are arranged in the chain:

A+B+A+B+A+B   (alternating)   **Figure 17.5** Polymerization structures.
A+A+A+A+B+B+B+B (block)
A+A+B+A+B+B+A+A (random)

Polymer Properties and Characteristics

There are a number of unique physical properties that are used to describe the properties and behavior of polymers beyond their chemical composition. One of the most important is the *glass transition temperature*, commonly represented as T_g. It would be similar to the melting point of a solid except that for polymeric materials, it is not usually at a specific temperature. The polymer can soften over a range of temperature and has a number of different shapes. This shape is a function of a number of factors including MW and MWD. T_m is the temperature at which the polymer has totally liquefied. A laboratory device is used to melt the polymer, and its softening measured against a reference point. An output graph would look like Figure 17.6:

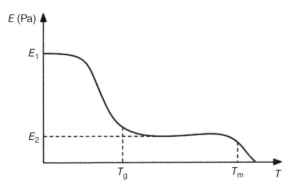

Figure 17.6 Differential scanning calorimetry (DSC) graph: energy (Pa) versus temperature.

   With these properties in mind, we can describe the general nature of the various types of polymers as follows:

1) Thermoplastics. This refers to a polymer that can be melted into a liquid phase and then cooled to reform the original solid polymer. For most thermoplastics, this cycle can be repeated numerous times; however, over a long period of time, some degradation in the MW of the polymer will be seen (this is one of the limitations of plastics recycling where the properties of the recycled polymer may limit its potential uses to less valuable products than the original polymer). The T_g would typically be at room temperature or above. Polyethylene, polypropylene, and polystyrene are examples of such materials and find use in molded toys, food wrap films, and molded shapes of many types.

2) Thermosets. These types of polymers are chemically cross-linked in such a way that they do not have the ability to "melt" in the traditional sense. Epoxy resins are examples of such materials. When heated, they will tend to char and burn. They are suitable for rigid plastic structures such as tanks. Because they cannot be melted in a normal sense, it is not possible to recycle these types of plastics in an economical way.

3) Elastomers. These are polymers whose T_g is below room temperature and thus can be stretched or fitted around a mold at room temperature. A simple rubber band is an example of such a material. These materials find use in applications where flexibility at room temperature is important, such as gaskets or medical tubing.

4) Emulsions and Latexes. These are suspensions of polymers in either an aqueous or organic liquid where the polymer is suspended in the solution by means of surface chemistry which interacts with the solution. A household latex paint is an example of such a system. The paint may look as if it is a solution of polymer in water, but if a paint can is left unused, but covered, for an extended period of time and then opened, a clear water layer is seen at the top. Agitation will re-disperse the latex polymer in the water solution. This behavior is enabled via a unique polymerization technology that coats the surface of the polymer beads with functional groups that are attracted to polar molecules such as water.

5) Engineering thermoplastics. This general term is used primarily to designate thermoplastics with very high temperature resistance (i.e., high T_g) and/or having a high degree of chemical resistance to solvents and acids. Examples of such materials include nylon (frequently used in "under the hood" automobile applications where gasoline resistance is critical), polycarbonate (having high temperature resistance and high impact strength), acrylonitrile–butadiene–styrene terpolymer (ABS), and fluorinated polymers.

Polymer Processes

The polymerization of monomers has the same aspects of kinetics as discussed in Chapter 4. However, in polymerization we have several different reactions occurring simultaneously: First, the rate of the polymer chain growth, second, the rate of chain growth termination, and finally, the rate of branching of the polymer chains.

Monomers can be combined to obtain polymers, with the aforementioned characteristics in various ways as follows:

1) Thermal. The double bonds of many monomers, such as ethylene or styrene, will be activated by thermal energy and can be polymerized through simple thermal processes. There is usually little control over the characteristics of the polymer produced. This was the primary process for polymerization of compounds such as styrene and methyl methacrylate in the 1940s and 1950s prior to the discovery of catalytic processes.

2) Catalytic. The use of catalysts to polymerize monomers has been used for many decades, allowing the control of chain length (MW) and branching. These processes can be either liquid phase (under pressure in liquid monomer) or vapor phase, producing a particulate product. Vapor-phase and liquid-phase catalytic processes will produce different products in terms of density and other physical properties. Until recently these processes were required to be shut down when the catalyst activity diminished below an acceptable point. Recent developments in polymerization catalysis have produced catalysts that are so active that the amount required does not justify recovery and they can be left in the polymer product as a trace residue, which does not affect its performance or properties. These types of processes can be run in the liquid phase (frequently under pressure due to the low boiling points of many monomers) or in the gas phase.

3) Condensation. This refers to a copolymerization involving not only joining two different monomers but also generating a chemical by-product such as water or HCl in this process. The polymerization of amides and amine functional groups in various ratios to produce different forms of nylon is such a process as shown in Figure 17.7.

The process of producing polycarbonates via reacting phosgene and bis-phenol is another common process of this nature.

In condensation polymerizations, a by-product produced as part of the coupling reaction (H_2O or HCl) must be removed from the process, complicating the process from a chemical engineering standpoint, especially since the polymer streams are high viscosity.

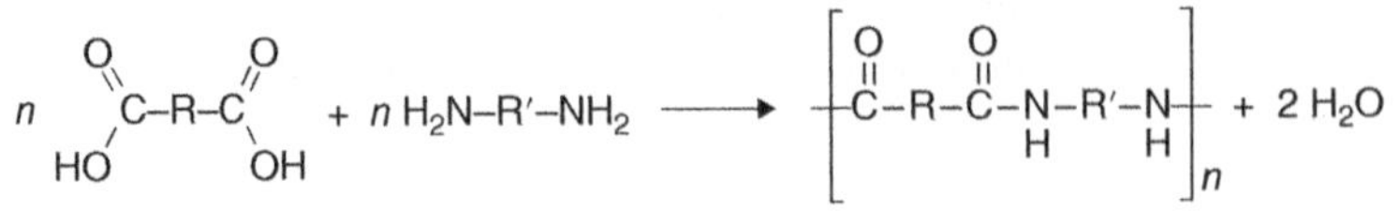

Figure 17.7 Formation of nylon from an amide and carboxylic acid.

4) Copolymerization. If we want to copolymerize two or more different monomers, we can use heat and free radical activation to produce some combination of polymers A and B (and C, D, etc.). Examples of such materials would
   include styrene–butadiene rubber (SBR) and ABS. These reactions also
   couple two or more different monomers, but as opposed to condensation
   polymerization, they do not split out a third molecule.

   Suspension polymers, such as latexes, are typically produced in the liquid
   phase, in a sequential batch process that builds layers of the desired polymer
   composition, and then in a final layer including a "water-loving" molecule
   that attracts water and allows these polymers to be suspended in water.

5) Chiral and Tactic Polymerizations. The carbon atom is somewhat unique in
   that it does not have a "linear" geometric center. The carbon atom has four
   bonds, which are in the shape of a pyramid and two identical chemical compounds (by chemical formula) and can be represented by two different structures. This is the case with the amino acid alanine as shown in Figure 17.8.

Figure 17.8 Chiral molecule illustration.

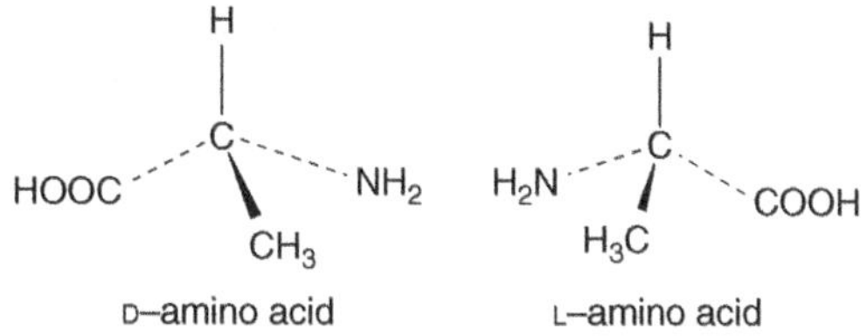

D–amino acid   L–amino acid

These two structures are not superimposable on each other. One, as seen in
Figure 17.8, is "left handed" while the other is "right handed." If a polymerization involves a monomer with such a chiral center and a double bond, the
polymers manufactured will generally have very different properties.

If we use the generic symbol "R" to represent a functional group, it is possible
to produce a polymer with the "R" being randomly distributed by position
(called *syndiotactic* polymers), virtually all on one side of the polymer chain
(called *isotactic* polymers), or randomly distributed (*atactic* polymers). These
polymer structures would be seen in Figure 17.9.

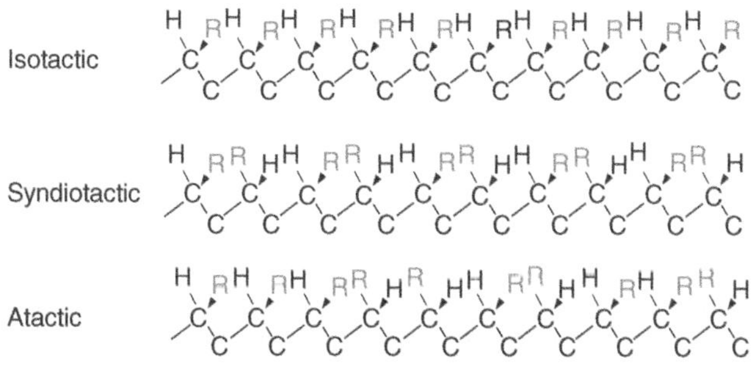

Isotactic

Syndiotactic

Atactic

Figure 17.9 Polymer "tacticity" structure.

If these three polymers were polybutadiene, the differences in T_g and physical properties are significant. One can be hard enough to use as a golf ball cover and another a low T_g material with no practical use. These polymers are examples of *chiral* polymers. Figure 17.10 shows this difference in the polymerization of isoprene monomer, with the methyl groups either alternating or totally atactic.

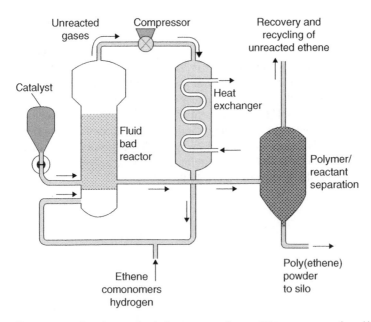

Figure 17.10 *trans* versus *cis* isomers.

Gutta percha (*trans*-1,4-polyisoprene)

Natural rubber (*cis*-1,4-polyisoprene)

The "Gutta" version is hard and rigid while natural rubber is soft and stretchy. One of the processes for the manufacture of polyethylene is shown in Figure 17.11.

Unreacted gases

Compressor

Recovery and recycling of unreacted ethene

Catalyst

Heat exchanger

Fluid bad reactor

Polymer/ reactant separation

Ethene comonomers hydrogen

Poly(ethene) powder to silo

Figure 17.11 Gas phase polyethylene process. Source: Diagram was produced by the Centre for Industry Education Collaboration, a non-profit organization and an integral part of the Department of Chemistry, University of York, UK. http://www.essentialchemicalindustry.org/

This is what is known as a catalyzed gas-phase reaction system. The polymerization reaction itself is exothermic and occurs in a fluidized bed. A heat exchanger is used to recover the heat of the exothermic reaction and use some of it to preheat the inlet ethane (ethylene) feed gas. The polymer product is separated from the unreacted ethylene/ethane and recycled. The final product is in a powdered form and is sent to storage silos. We can see many unit operations, which we have already discussed, used in this process, including reaction engineering and kinetics, thermodynamics, heat transfer, solids handling, and liquid–solid separation.

Polymer Additives

Very few polymers are used in the "as produced" condition. In the case of thermoplastics, it is common for the end product of manufacturing to be a small plastic pellet, which is shipped to the customer in bags, boxes, or bulk rail cars. The pellet is produced from a high temperature process stream or molded from fine particles produced from a vapor-phase reaction.

As the end user produces a final product (i.e., piping, tubing, molded shell, or building siding), colorants, flow processing aids, viscosity modifiers, and other additives are added to the molten polymer stream prior to its molding into a final useful product. The final polymer properties must match the final end-use demands in terms of temperature and solvent resistance, color, resistance to UV light, and abrasion.

End-Use Polymer Processing

The type of process used to fabricate the final product will be primarily determined by its end use and can include blow molding, extrusion, injection molding, or fiber spinning in the case of such materials as nylon and polyesters. The unique chemical engineering challenges in designing such process equipment include the typical high viscosity of molten polymers, which in turn reduces heat and mass transfer rates, resulting in long times to heat and cool and long times to mix and disperse materials.

The formation of polymers into final products is done in numerous ways, depending upon the application and properties of the starting polymer.

1) Extrusion. In this process, the polymer (or copolymer) is melted and then pushed through ("extruded") a die, which contains the form of the desired product. Plastic shapes, sheets, and toys would be produced by such a process.
2) Blow Molding. This is the type of process used to make plastic bottles. The polymer is extruded as a rod and then inserted into a die. A gas

stream is then injected into the rod, causing it to expand into the die shape, leaving the inside hollow.

3) Injection Molding. In this case, the entire cavity of a mold is filled with a molten polymer, and after cooling, is rejected and the process repeated.

In all of the aforementioned processes, the thermal and flow properties discussed are critical in designing the correct type of process and specifying the optimal processing equipment.

4) Spinning. In this case a polymer is extruded into a very fine diameter rod and these "fibers" are spun together into what amounts to a multicomponent thread wound up on to a spool. These threads (typically nylon, polyester, etc.) are ultimately unwound and spun into clothing.

As opposed to specific chemical entities, polymers and multicomponent polymers will have a broad range of properties, and many product demands can be met by several different polymer systems and blends. In any case, the polymer properties must match what is required in the final end use. Examples include impact strength, tensile strength, clarity, glass transition temperature, and others. Specific chemicals such as acetone have a specific boiling point, freezing point, vapor pressure, viscosity, density, and surface tension, which are independent of the process by which they are made. This is not the case with polymers and makes polymer processing much more empirical in nature.

Plastics Recycling

There is much activity in this area directed at trying to reduce the volume of plastics in landfills. There are several challenges in accomplishing the goal of total plastics recycling as follows:

1) Polymer Degradation. Polymers are subject to degradation in terms of their MW and physical properties each time they are melted and reprocessed. We can see this in commercial products made from recycled plastics when the label says "made with 10, 20, etc. % recycled plastics." Few materials can be reprocessed to their original condition due to degradation and the contamination of one plastic with another, not allowing accurate process control to achieve the needed physical properties.

2) Additives. The color and physical property modifiers added for the original end use may or may not be suitable for the recycled use, thus eliminating all but the least severe applications such as low quality plastic garbage bags. The cost of separating the various polymers and additives to the degree necessary is currently much too cost prohibitive for commercial use, with some sort of subsidy or tax credit.

3) Depolymerization or Pyrolysis. In this form of plastics recycling, the plastics are "burned" in an anaerobic system in the absence of oxygen. In this

type of process, the polymer decomposes back into its starting monomers (i.e., ethylene, styrene, etc.), which are then separated via conventional chemical engineering separation techniques such as distillation or absorption. This type of process provides much more flexibility in the use of the recycled monomers.

Summary

The chemical engineering issues with polymer manufacture, processing, and end use are more complicated than traditional chemical processing. The properties of polymers are orders of magnitude different from traditional chemicals, impacting fluid flow, heat transfer, and reaction engineering. In addition, end-use properties (in the eye of the end user) can be achieved by many different types of polymers, including additives. The product and process are mutually connected far more than in the case of traditional chemicals and chemical engineering. The use of catalysts to produce many different forms and types of polymers is also unique challenge.

Coffee Brewing and Polymers

The bag in which much coffee is packaged is usually a foil-lined (to prevent oxygen and water from entering) plastic bag, which must be strong enough to withstand normal handling in the grocery store, distribution channel, and home. It must not rupture when dropped from reasonable heights. This value can be measured in lab tests. The materials used in containing the coffee, including the new "pods" used in single serving systems, must have a high enough T_g to withstand the temperature of the brewing water and not melt in the coffee brewer or in the dishwasher.

If your coffee cup is a foamed polystyrene cup, do you recycle it?

Discussion Questions

1   How many different types of polymerization processes are run at your location? Is it clearly understood why a particular process is used to produce a particular product?

2   For the polymerization processes to run, are their limitations clearly understood in terms of MW, MWD, tensile strength, etc.?

3   How are the polymers formulated? With what type of equipment? How was it chosen? Are you prepared for possible customer requests for changes?

4 What possible new products could be produced in your current equipment?

5 How well is the heat transfer in and out of the polymer streams understood? What possible improvements are possible?

6 Are you familiar with all the possible crystallinity and density forms of your products and how they could be changed?

Review Questions (Answers in Appendix with Explanations)

1 Polymers are long chains of:
  A __Polys
  B __Mers
  C __Various mixtures of polys and mers
  D __Monomers

2 Latex polymers are unique in that they:
  A __Are used in tennis shoes
  B __Are stretchable
  C __Are suspensions of polymers in solution
  D __Are combinations of thermoplastic and thermoset polymers

3 The uniqueness of elastomers is that they:
  A __Have a T_g below room temperature
  B __Have a T_g at room temperature
  C __Have a T_g above room temperature
  D __Have a T_g that is controllable

4 *cis-* and *trans*-isomers differ in:
  A __Where functional groups are positioned on the monomer backbone
  B __Their preference for being *cis* or *trans*
  C __Their ability to change positions
  D __Cost

5 "Condensation" polymerizations are unique in that they:
  A __Produce polymers that condense when it is cold
  B __Split out a molecule during the polymerization process
  C __Prevent another monomer from entering the process
  D __Provide a barrier to another polymerization process occurring

6 Polymer additives can be used to affect or change:
  A __Color
  B __Flowability
  C __Ability to foam
  D __All of the above

7 A differential scanning calorimetry (DSC) tells us:
  A __Softening temperature range of a polymer
  B __Softening temperature ranges within a polymer
  C __Degree of crystallinity within a polymer
  D __All of the above

8 Chemical engineering challenges presented in processing polymers include all but:
  A __High viscosity
  B __Slow heat transfer rates
  C __Knowledge of the polymers being processed
  D __Difficulty in mixing

9 The technical challenges in recycling of plastics and polymers can be affected by all but:
  A __Purity of plastics and ability to separate into various types
  B __Energy value of plastics
  C __Legislation
  D __Landfill availability

Additional Resources

Sharpe, P. (2015) "Making Plastics: From Monomer to Polymer" *Chemical Engineering Progress*, 9, pp. 24–29.
Stein, H. (2014) "Understanding Polymer Weld Morphologies" *Chemical Engineering Progress*, 7, p. 20.
Villa, C.; Dhodapkar, S.; and Jain, P. (2016) "Designing Polymerization Reaction Systems" *Chemical Engineering Progress*, 2, pp. 44–54.
https://www.usm.edu/polymer.

18

Process Control

It almost goes without saying that a chemical process needs to be controlled. But why? This is an important question to answer prior to deciding *how* we want to control the process. Some typical reasons include the following:

1) Internal Process Disturbances and Changes. These can take many forms including changes in feed rates, temperatures, pressures, and compositions of incoming process streams. Examples include cooling water temperature and pressure, steam temperature and pressure, and changes in air pressure and electrical voltage used to control process equipment.

2) External Conditions. These would include weather (primarily external temperature and weather conditions such as wind, rain, and humidity) and changes in items mentioned in #1 earlier if the sources of the variations are outside the control of the process operation. If a feedstock or raw material being used is coming from a source outside the organization's direct control (feedstock arriving in a tank truck, tank car, or pipeline), there must be a process in place to ensure that the composition and quality of this material is what is specified.

3) Safety and Environmental Impacts and Regulations. Many processes, if not under control, could emit hazardous materials that could injure the people within the plant's perimeters or in the surrounding communities. Most chemical process operations also operate under several kinds of operation permits issued by local, state, or federal authorities. These permits frequently limit environmental discharges to the air and water. Any process control strategy must assist in complying with such regulations. From a safety standpoint, the means by which a process is controlled *must* ensure that operating personnel are not subject to situations where their safety and lives are threatened. The reactive chemicals area discussed earlier is one of these areas.

4) Planned Changes. In many chemical operations, the "normal" process operations may be deliberately changed as a function of customer demand,

Chemical Engineering for Non-Chemical Engineers, First Edition. Jack Hipple.
© 2017 American Institute of Chemical Engineers, Inc. Published 2017 by John Wiley & Sons, Inc.

regulatory requirements, or external conditions. Examples here would include changing gasoline composition due to differences in volatility requirements as the seasons change, especially in geographic areas where winter and summer weather conditions are quite different, and specialty chemical operations to reflect changes in product composition or changes in ratios or temperatures within a semi-continuous operation.

Other reasons we desire to document, as well as control chemical operations, include the following:

1) Regulatory and Process Documentation. For example, when a pharmaceutical is registered by the Food and Drug Administration (FDA) in the United States, it is necessary to document not only the final product composition but also the process variables and conditions used to produce the product, as they are tied directly to the FDA registration approval process.
2) Product Quality Issues. Many customers using chemical products have defined specifications for their products, which must be met; however, in some cases the performance of the product in the end-use applications has not been defined adequately by measurable items. In some of these cases, the process operating conditions can be used to provide information to better control a process for a particular customer to partially substitute for end-product analytical information.
3) Process Learning. There is no chemical process operating today where everything is known about how all variables will respond to all possible changes. It is common for a newly built plant to have instrumentation, analytical devices, corrosion coupons, and other such items installed within it for the simple reason of data collection to learn more about the process to improve understanding and how it might be better monitored and controlled in the future. This information is analyzed offline for its learning value, but is typically not used to control the current plant operation.

Elements of a Process Control System

We first need to have a desired value for something we are measuring. This could be temperature, level, mass weight, flow rate, composition, or pressure. This has usually been determined by previous experience, process knowledge, and input from customers. (Customers in this context could mean the next unit process within a larger process run by the same company.) As a simple example, consider the human body temperature. We know the normal desired number for this is 98.6°F.

The second part of the system is a way to measure what is of concern to us. This could be a thermocouple or RTD device for temperature, a differential bubbling device for level, an orifice plate for flow, an online (or offline) analytical device to measure composition, or a diaphragm type of device to measure pressure. At home, we do this with a thermometer, inserted under the tongue and held there until the temperature reading does not change. This allows us to know whether our body temperature is at the "set point."

The third part of the system is the way to connect these two parts. In other words, what do we do if the measurement is not what it needs to be? We need an action of some sort to make these two measurements to be the same. We refer to this as a control action. This could be a change in heating or cooling to control temperature, an increase or decrease in flow rate or pump speed to control flow or level, or a change in one feed rate relative to another to change composition. Another peripheral question is to ask what are the consequences of this variable not being what it is supposed to be? To what degree? And for how long? The answer to these last two questions will have a great impact on the nature of the response (what we will now call the *control action*). In our simple example, it might be to take an aspirin tablet, drink something cold, or call the doctor (or possibly some combination of these three actions). In a chemical process, this action could be either manual or automatic, depending on the seriousness of the deviation and the speed of the response necessary.

In our human body temperature case, a minor deviation from normal body temperature might just call for taking two aspirin tablets and getting some extra rest and fluid consumption. If, however, the measurement rises to 103°F, we now are concerned about something far more serious such as the flu or pneumonia. We take more serious action such as seeing the doctor or going to an emergency room. Our response is *proportional* to the deviation from normal.

1) Measurement. First of all, we need a way to measure what is of concern to us, for example, level, pressure, or composition. The exact way we do this will be a function of the desired accuracy and response time of the measurement system (how long does it take for the measurement device to detect a change and transmit it to the part of the system, which will analyze and react to the measurement?). If we are measuring pressure, differential pressure instruments are typically used. If direct contact with the fluid is not possible, then various types of electronic or ultrasonic devices may be used. In some cases, depending on the nature of the fluid, magnetic property differences can be used to measure the level external to the tank, avoiding contact. When using differential pressure as a measurement approach, it is important to take into account any density changes in the fluid due to compositional or temperature changes. A bubbling device responds to pressure of a height of liquid, and if the density of the fluid is increased, a change in density will be seen as a change in level. It is also important to remember that any gas introduced to measure level must be able to be vented from the process in a safe way and not contribute to flammability or environmental concerns.

Temperatures are typically measured by thermocouple or RTD devices, which use the difference in two different metals' change in conductivity or electrical resistance as a function of temperature.

Flow measurements are typically made with a device that temporarily restricts the flow of fluid through a hole smaller than the diameter of the pipe. These devices were reviewed in Chapter 5.

2) Comparing and Evaluating. After a measurement is made, we need to decide what we should do in response. What is the difference between what we observe and what we need? How fast is the change occurring? What are the consequences if a change is not made? How fast do we need to respond? Under what conditions? This is sometimes written as E (error) = MV (measured value) – SP (set point).

3) Responding. After we have measured or observed and compared what we see or measure to what we need, we need to decide how to respond to the difference. This can be done in a myriad of ways depending on the nature of the difference (error) we see, the necessary speed of response, and the nature of the response. The most often seen aspect of this last part of the cycle is some kind of control valve that regulates flow, heating or cooling, pressure, or composition. These valves have many different types of design and can be designed and installed in many different ways, which we will discuss later. This cycle of measuring, comparing/evaluating, and responding is what we call a *control loop*.

Control Loops

This cycle of measure, evaluate, and take action (or respond) is commonly referred to as a control loop. We continuously measure, evaluate, and respond. The speed at which this is done can be anywhere from many times per second to just several times an hour, depending on the rate of change in the operation being controlled, how critical it is to maintain precise control, and the safety/environmental implications of loss of control. In addition to the simple human thermometer example already discussed, how a turkey is prepared for Thanksgiving is another analogous situation we can all relate to. We have a cookbook that says if the turkey weighs a certain amount and is stuffed, then it should take so many hours to bake at a given temperature range. We prepare the turkey and for the first several hours, we pay little attention. As we get near the end of the cycle, we start to inspect the skin and poke the turkey to "fine-tune" the final cooking temperature. The speed of the control loop is very slow at the beginning and speeds up at the end. The analogy in a chemical reactor might be a slow reaction producing an intermediate, which then reacts very fast with a newly added material to produce a finished product.

There are numerous types of control loops that can be designed and implemented, and with today's advanced measurement, calculation, and computer capabilities we have, it is possible to design an infinite number and types of control loops and systems. We will cover only the most basic concepts here.

On–off Control

This type of control is analogous to the type of control system found in most home heating systems. First of all, a temperature is set. If the temperature deviates from this temperature, the fan and heating/cooling system turns on until the desired temperature is reached and then the system elements shut off. An industrial analogy might be maintaining the temperature of tank contents at a certain temperature with a heating or cooling coil. Another might be controlling level in a tank and turning a pump on or off based on the level measurement. This type of control will always have the possibility of not exactly reaching the desired temperature due to simple overshoot or undershoot based on system volume, lag time, etc. If offset is tolerable and the system is not demanding in some way, this may be an acceptable control approach. For a tank level on–off system, with pumping in and out, we might see a control room graph that looks something like that shown in Figure 18.1.

Figure 18.1 On–off control for tank level.

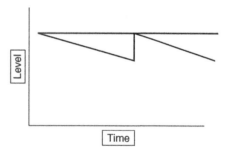

This would be similar to what occurs in a home heating system, except that we typically do not see the graph. The manufacturer of the thermostat has preset (which might be adjustable by the homeowner) a certain deviation, at which point the gas flame is ignited by the pilot light and the blower starts. The small differences in the actual temperature in the house might be less than 0.5°F, too small for anyone to notice. Since sometimes the error is high and sometimes it is low, the average of the error is acceptable. The same logic applies to the start of an air conditioning compressor or a sump pump in the basement of a house. This type of control will always have some level of error that is not automatically compensated for since there is no way to integrate the total error over time. In the case of small variations in a tank level or a small degree of temperature change in a home, these differences average out over time in a way that is not noticeable from a practical standpoint. This type of control may put process equipment under abnormal stress as the final control element (pump, heater, compressor, etc.) is starting and stopping on an irregular basis as opposed to running continuously.

Proportional Control

This type of control includes a response to the error that is proportional to the amount of the error. In our home heating example, the fan speed could be increased (requiring a variable speed motor) if the temperature was much higher or lower than the normal. In our tank system the output of the pump (which is now running continuously and has a control valve throttling its output) will be changed in response to the *amount* of error, not just the fact that there is an error. Mathematically, we would describe this type of control as follows:

$$CE = K_c e + m$$

where CE, the position of the control element, is proportional to the amount of error plus the reset value; K_c is the controller gain (sensitivity is one way to look at this or how fast and how much the controller reacts to an error); and m, the reset value, is the position of the control element when the process is at the set point. We need to ensure that we understand the value of "m." At any point in time, a control valve has a position, "m." When asked to change, its position moves according to the aforementioned equation, proportional to the amount of error and the speed with which we desire a response (K_c), but the initial valve setting, "m," is not changed. Without a change in this value, we have a permanent offset in the valve position versus what we want. This produces an "offset." A graphical display of this response would look like Figure 18.2, if displayed on a control graph.

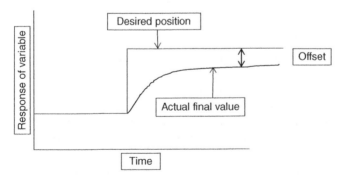

Figure 18.2 Response of proportional control: offset.

Though a home heating or cooling system is normally not a proportional system, it will typically display a response such as this, and the reason we don't notice is that on occasion it is above and in others below (depending on where the system starts from) and the small difference and the average temperature in the room being close enough for us to feel comfortable. Whether an offset is tolerable in a chemical process, control loop is quite another matter. We need to think carefully about the consequences of the variable of concern not being where it is desired to be and for how long.

Proportional–Integral Control

In this type of control, the error around the desired set point is constantly evaluated and the valve position is changed until the desired value of the controlled value is reached. In effect, "m" is constantly adjusted until the desired set point is reached. A graph of this type of control would look something like that shown in Figure 18.3.

Figure 18.3 Control response with integral control.

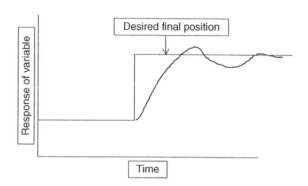

There will be both an "overshoot" and "undershoot" as the system integrates the error until the desired final control point is reached. The primary advantage of this type of control loop is that it will eventually reach the truly desired set point. However, there will be an overshoot and undershoot on the way to the final steady state, and the implications of this must be considered. There are two parameter choices in this type of control. First is the gain (K_c), the same as discussed for simple proportional control (how much reaction is there to a given amount of error?). The second design aspect of this type of control is what we call the "reset" time. In other words, how often does the control system measurement check to see if it has reached its final desired point and whether it is over or under that desired set point? This could be milliseconds to minutes, depending on the nature of the system being controlled and its response to change.

We can view two extremes of these settings as seen in Figure 18.4.

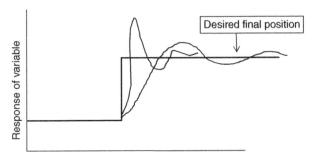

Figure 18.4 Impact of choice of integral reset time.

Line "A" would represent a high K_c and short reset time, while line "B" would represent a lower "K" and a slower reset time. The choice of this setting will depend on the impact of overshoot, the amount of time where not being at the final set point is tolerable, the impact of these controlled variables on downstream operations, and the nature of the controlled variable itself and what affects its change. In a typical plant start-up, the gain and reset time would be estimated based on process knowledge and typically programmed into a process control computer or set manually in a simple controller. Based on actual process behavior and response, these settings would be adjusted during a plant start-up.

Derivative Control

This additional level of sophistication can be added to either of the aforementioned strategies. In derivative control, we assume that we have enough knowledge about the process that we can estimate, ahead of time, how the process will respond to a known change. For example, if we are running a reaction that has a known ratio of "A" to "B" and we increase the rate of "B," we know that if we increase "B" by 50%, the rate of "A" should also be increased by 50%. There should be no need to have a downstream analysis output telling us that the amount of "A" should also be increased by 50%. A more complex illustration of this strategy would be changing cooling or heating to a reactor based on a change in feed rates or temperatures of feeds to a reactor based on our knowledge of heats of reactions, heat capacities of materials, etc. There is always some uncertainty in this type of control as it assumes that the effects of upstream changes in process variables are all known and can be programmed into a process control scheme. This is rare, but the need for a set point to be ultimately achieved is often critical, and the result is that the large majority of chemical process control loops are based on a proportional–integral (PI) scheme.

We can view a summary of these approaches to control in Figure 18.5.

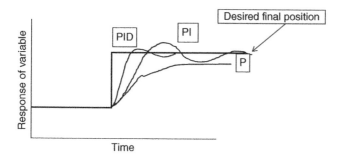

Figure 18.5 Response of various control system types.

Both PI and PID control systems will eventually reach the desired set point, and both will overshoot the desired set point to some extent. The control loop settings that will determine to what extent this happens will be fine-tuned during a plant start-up.

Other factors that must be considered in the design of a control system include "dead time," meaning the time difference between when the variable is measured and when response to that measurement is made. If this time differential is high, the response of the control system to changes or errors may become out of phase with the dynamics of the process. The same kind of result is possible if the time for the actual measurement is much slower than the process response to change.

There are many sophisticated mathematical modeling techniques for determining the various control scheme constants, including what is known as Ziegler–Nichols, allowing tuning of a control loop based on empirical process information.

Other types of process control structural design include the following:

Ratio Control

If we know that a change in one feed rate should be accompanied by a known change in another variable, we can make this change at the same time. It is important that we understand all the impacts of a feed change, including heat release in an exothermic reaction or gas production in a reaction.

Cascade Control

We can think of this type of control as "embedded." For example, if we have a very slow chemical reaction system providing the feed to another, a much faster reaction system, the control loops for these systems would be very different as one is feeding the other and their rates are very different.

Measurement Systems

In order for a variable to be controlled, it must be measured and that measurement information must be used in the control loop design.

Flow measurements were discussed previously in the unit on fluids and pumps.

Pressure measurements are most frequently done by a mechanism that senses differential pressure against a reference point. A diaphragm device outputting an electronic signal can serve this purpose.

Temperature measurements are done with thermocouples or with RTDs, where resistance changes between two different metals generate an electronic signal.

Level measurements can be done through the use of differential pressure measurements, magnetic resonance, or ultrasonic measurements.

Control Valves

The final element in a control loop is the control valve. This valve is actuated via air or an electronic signal and moved to varying degrees depending on the difference in measurement between the actual versus the desired value of a particular measurement. There are many choices in such valves, and the choice needs to be made based on the desired response we want as a function of the amount of error measurement and how fast the response needs to be. Basically, control valves are characterized in two ways as follows:

1) Size. This relates to the maximum flow that a given valve will allow. This will vary as a function of pressure drop available and fluid properties, as discussed earlier.
2) Flow versus % Opening Response. This is basically a plot of percentage of maximum flow (which will vary with fluid properties already discussed) as a function of what % open the valve is. This is not necessarily a linear function as the geometric design of internals of the valve can greatly affect this relationship and these differences are a major part of the valve choice decision.

Though there are many more possibilities based on internal geometric valve design, we can classify many valves as linear, fast opening, or equal percentage as shown in Figure 18.6.

It is also possible to design a valve with a more complicated response curve if desired. A globe valve is illustrated in Figure 18.7.

We can see that a very small adjustment in the valve stem will allow a great deal of flow, and this type of valve would typically be characterized as a fast opening valve. "Trim" is the word used to describe the degree of flow response

Figure 18.6 Flow versus % valve opening. Source: Reproduced with permission of Engineering Toolbox.

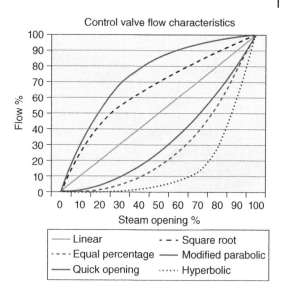

Figure 18.7 Elements of a globe valve. Source: Padleckas, https://commons.wikimedia.org/wiki/File:Valve_cross-section.PNG. Used under CC BY-SA 3.0, https://creativecommons.org/licenses/by-sa/3.0/deed.en. © Wikipedia.

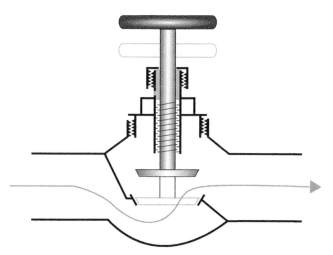

to valve opening. The geometry of this valve trim can be designed to give almost any response as shown in Figure 18.6.

An equal percentage valve has an internal design that very slowly allows flow versus its physical opening (see Figure 18.6). These valves are excellent for close control over a broad range of conditions.

A ball valve is shown in Figure 18.8.

Figure 18.8 Ball valve. Source: Castelnuovo, https://commons.wikimedia.org/wiki/
File:Seccion_valvula_de_bola.jpg. Used under CC BY-SA 4.0, https://creativecommons.org/
licenses/by-sa/4.0/. © Wikipedia.

A quarter turn will move the valve from no flow to 100% flow. This type of valve will produce a more linear response curve.

A gate valve is one where a simple linear plate is installed perpendicular to the flow and the valve moved up and down as desired, as seen in Figure 18.9.

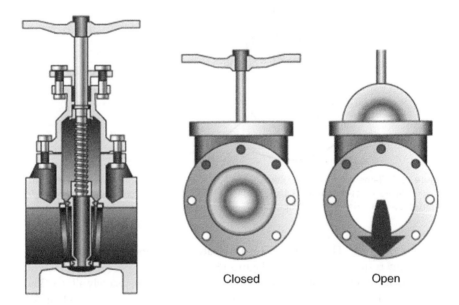

Closed   Open

Figure 18.9 Gate valve. Source: Reproduced with permission of Google.

This type of valve will typically have a linear response curve. A butterfly valve is shown in Figure 18.10.

Figure 18.10 Butterfly valve. Source: Qurren. https://commons.wikimedia.org/wiki/File:Yagisawa_power_station_inlet_valve.jpg. Used under CC BY-SA 3.0, https://creativecommons.org/licenses/by-sa/3.0/deed.en. © Wikipedia.

Because of their reliance on sealing against an interior surface, it is difficult to ensure 100% shut off in this type of valve.

One other type of "on–off"-type valve sometimes is a check valve, which is intended to prevent flow in a particular direction primarily due to safety or quality concerns. This can be done via a flap that is designed to be totally open in one direction and allow no flow in the other direction (see Figure 18.11).

Figure 18.11 Check valve with flow indication.

When this type of valve is installed, it is critical to ensure that the arrow (indicating the direction that flow is allowed) is in the correct direction. Obviously, if such a valve were to be installed in reverse, the safety or environmental consequences could be severe, allowing the flow to go in the exact opposite direction vs. intended.

Valve Capacity

The capacity of a valve is normally given by its C_v or capacity of water flow at 1 psi pressure drop with the water at 60°F. The higher the C_v of a valve, the greater its capacity. This provides a standardized way of comparing different valves in the same service and can assist in choosing a valve when pressure drop is a prime consideration. For example, if a particular valve had a C_v of 12 versus a C_v of another valve at 4, we would expect three times the flow at the same fluid conditions and pressure drop across the valve. When doing these calculations or comparisons, remember to maintain the same set of units.

Table 18.1 shows the relative capacities of different valve sizes.

Utility Failure

Since the movement of control valves is driven typically by electrical or mechanical energy, it is important to decide, prior to installation and start-up, how a particular valve should respond if the utility driving its change is lost.

Table 18.1 Relative valve capacity.

	Control valve flow capacities (Cv*)						
	Valve size (in.)						
Valve type	0.75	1.5	2	4	6	8	10
Single-seat globe	6	26	46	184	414	736	1150
Double-seat globe	7	27	48	192	432	768	1200
Sliding gate	5	20	36	144	324	576	900
Single-seat Y	11	43	76	304	684	1216	1900
Throttling ball	14	56	100	400	900	1600	2500
Single-seat angle	15	59	104	416	936	1664	2600
90%-open butterfly	18	72	128	512	1152	2048	3200

Source: Chemical Engineering Progress, 3/16, pp. 51–58. Reproduced with
    permission of American Institute of Chemical Engineers.

When should a valve fail in the "open" position? Any time the valve is controlling cooling water or other utility that is designed to reduce the rate of an exothermic chemical reaction. The reaction will not stop just because utilities are lost. In fact, if it is a reactor vessel and the agitation is lost for the same reason, this will only increase the potential danger of the situation. A backup electrical supply may also be necessary to keep the process under control.

When should a valve fail in the "closed" position? If we are running an exothermic reaction, we would want all of the feed materials into the reactor to stop, in combination with the cooling water failing to open.

When might we want a valve to fail in its "last position?" If we are melting a heat-sensitive material, we might choose this option versus having the material freeze or decompose if exposed to a higher temperature.

It is possible to design a valve and utility system to do any of these aforementioned options. The important thing is to think about the ramifications of utility loss prior to designing and installing a control valve system.

Process Control as a Buffer

It is important to remember that process control is used not only to control a specific process variable but also to provide a buffer between unit operations within a process or plant. Examples of these include the following:

1) Inventory Management. A unit operation such as a reaction system may be providing the feed to a separation unit. Either rates may need to be slowed to take into account another's slower rate or downtime. Restrictions in raw material availability or product storage are also examples.

2) Utility Restrictions. If there is an outage at a utility, either on-site or off-site, there may be a need to slow a unit operation or change the conditions under which it operates.
3) Emergencies. We must always consider the possibility of utility loss. The process control system must be designed to react to this situation in a safe way. In many large plant situations, backup electrical generations units are designed to activate and keep a process's key units and controls operating for a specified amount of time.

Instruments that "Lie"

In many situations, an instrument's output is calculated based on physical properties that are assumed to be constant when this is not necessarily true. For example, a level instrument that is not calibrated and corrected by density of a fluid may show level decreasing when it is actually increasing. This will cause the control system to add liquid instead of the proper response. This was actually one key cause of the large Texas City fire and explosion in Texas City, TX, in 2005 (see Figure 18.12).

It is important in process control design and instrument selection that we consider all the variables that might affect the measurement we are reacting to, such as the following:

1) Temperature. What is the effect of a temperature change in the reading? This applies especially to measurements of level and density and especially, in the latter case, to gases.
2) Pressure. An instrument calibrated for one set of pressure conditions can give off the wrong output if this is not compensated for, especially for gases.
3) Composition. There is a tendency to take averages of composition measurements and rely on them to provide an output. The exact compositional outputs as a function of concentration need to be measured.
4) Density. How can compositional change affect this measurement?
5) Viscosity. Temperature, as well as composition, can affect this measurement greatly. Have these differences been taken into account?

Summary

Process control is a critical part of any chemical process, no matter its level of sophistication. The control logic and design of the control loops must be thoroughly thought out, and all variables that may affect measurements and how the control system will respond to them must be considered.

Sponsored by
CCPS
Supporters

 Process Safety Beacon

CCPS
CENTER FOR
CHEMICAL PROCESS SAFETY
20 YEARS
An AIChE Industry
Technology Alliance

Messages for Manufacturing Personnel
http://www.aiche.org/CCPS/Publications/Beacon/index.aspx

Instrumentation — Can you be fooled by it?   March 2007

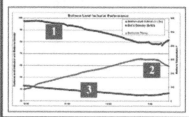

What happened?

A column was overfilled. However, before the incident, as shown in this instrument chart, the level *indication* in the bottom of the column (the dark blue line - 1) slowly decreased!

YES you can!

The level was measured with a displacement level indicator. Normally, when the displacer (green) is partially covered with liquid, it properly indicates level based on the changing force on the displacer as the liquid level changes (first and second drawings). But, on the day of the incident, the column was overfilled with cold liquid, completely submerging the displacer in cold liquid (third drawing). The level was above 100%, and the level indicator showed a high level alarm condition continuously. A high level alarm indicates an abnormal condition, and this should be an alert that something is not normal. In this incident, there was no response to the alarm condition.

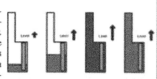

With the liquid completely covering the displacer, the instrument did not indicate liquid level. Instead, the force on the displacer gave a measure of the relative density of the displacer and the liquid in which it was submerged. In other words, the instrument was not designed to function properly if the level was high enough to completely submerge the displacer. The column was heated during the startup. As the temperature of the liquid increased (the green line in the graph above - line 2), the density of the liquid decreased (the purple line - 3). The change in density of the liquid changed the force on the displacer, resulting in a decrease in the "level" indication (fourth drawing, with hot liquid), even though the column level was actually increasing. The column overflowed, flammable material was released, and there was a major explosion and fire.

What you can do

Know what can fool you. Review examples of incidents where the instrumentation provided information that did not represent the data that was wanted (for example, density of the liquid, not level). This is not always an easy concept to grasp, so consult with the engineers and technicians who know the system best.

Understand how instrumentation works, and how it will respond to conditions outside the normal operating range, including, for example, control loops, venturis, orifice plates and impulse lines, differential pressure cells, level floats. Know whether instrumentation is normally energized, and the failure mode for valves, instruments and control loops following loss of pneumatic or electrical energy.

Know what you should be observing as part of normal operations, for example, balancing transfers into and out of equipment, changes in level. And, *NEVER* ignore alarms — find out what caused the alarm!

Understand whether components can be tested on line or whether an "out of service" test is required to confirm that an instrument is working.

PSID members use Free Search for "Instrumentation" or "Level Control."

Understand how your equipment works — and how it can fool you!

The Beacon is usually available in Arabic, Chinese, Dutch, English, French, German, Gujarati, Hebrew, Hindi, Italian, Japanese, Korean, Portuguese, Spanish, Swedish, and Thai. For more information, circle 103 on p. 63.

Figure 18.12 Understanding all variables that affect instrumentation response. Source: Chemical Engineering Progress and Center for Chemical Process Safety. Reproduced with permission of American Institute of Chemical Engineers and Chemical Process Safety.

Control of Coffee Brewing

Any coffee machine today has some kind of "control panel." The selection of coffee type, size of cup, and occasionally water temperature may be set points that a consumer can decide. The coffee drinker also decides how long the coffee is left on the hot plate or left in the vacuum container. The ratio of flavor additives and water type are also control points. The amount of mixing of the final product with cream and sugar or sweetener is also a control point. From a safety standpoint, most coffee makers are located in the kitchen, and building codes require them to use a ground fault interrupter (GFI) circuit, which will automatically cut power if any fluid comes in contact with the electrical circuit.

Discussion Questions

1   How are the various parts of your process controlled? How was this decided? Has anything changed about your raw material or products that would warrant a review?

2   Is it well understood why certain types of valves were chosen? Are the response curves available? If not, why not? When was the last time they were reviewed?

3   Are the control systems and hardware appropriate for the current list of concerns in the areas of safety and reactive chemicals?

4   Is it well understood what secondary variables could affect a critical process control loop?

5   Is it well understood how changes in process chemistry could affect process control?

Review Questions (Answers in Appendix with Explanations)

1   Proper process control is necessary because:
    A __Specifications must be met on products produced
    B __Environmental emissions must be within permitted limits
    C __Safety and reactive chemicals issues must be controlled
    D __All of the above

2   The elements of a process control loop include all but:
    A __Method to measure
    B __Manager's approval
    C __Way to evaluate the measurement versus what is desired
    D __Corrective action

3   Characteristics of a process which affect how it should be controlled include all but:
    A __Customer quality requirements
    B __Response time of measurements
    C __Degree of deviation permitted around the set point
    D __The mood of the process operator that day

4   The least sophisticated process control strategy is:
    A __On–off
    B __Off–on
    C __Off sometimes, on other times
    D __On–on, off–off

5   Integral control has a key advantage in that it:
    A __Has the capability to integrate
    B __Will eventually result in the process reaching its desired set point
    C __Oscillates around the desired set point
    D __Oscillates in a controlled manner around the desired set point

6   Derivative control allows a control system to:
    A __Anticipate a change in process based on input changes
    B __Preplan batch operations
    C __Provide a more uniform break structure for process operators
    D __React sooner to post process changes

7   If a process has a slow reaction rate process feeding into a very fast final reaction process, the type of process control likely to be used is:
    A __Proportional
    B __Follow the leader
    C __Wait 'til it tells me
    D __Cascade

8   Control valves can be characterized, in terms of their process response, by all but:
    A __Capacity
    B __Speed of response

C __Type of response
D __Materials of construction

9 A control valve curve plots:
A __% Open versus % closed
B __% Closed versus flow rate
C __% Flow versus % open
D __Flow rate versus air pressure supplied

10 A water cooling control valve, with loss of utilities, should fail to open if:
A __An endothermic reaction is being run
B __An exothermic reaction is being run
C __The utility water rates are temporarily dropping
D __There is no mechanic available to close

11 A control valve, with loss of utilities, should fail to close if:
A __An endothermic reaction is being run
B __It controls feed to an exothermic reaction
C __There is one person strong enough to close it
D __There is no other option

12 A control room variable may not indicate the actual process conditions if:
A __The sensor has failed or been disconnected
B __A physical property effect has not been taken into account
C __The operator is not looking at the correct screen
D __Any of the above

Additional Resources

Bishop, T.; Chapeaux, M.; Liyakat, J.; Nair, K. and Patel, S. (2002) "Ease Control Valve Selection" *Chemical Engineering Progress*, **11**, pp. 52–56.

Gordon, B. (2009) "Valves 101: Types, Materials, Selection" *Chemical Engineering Progress*, **3**, pp. 42–45.

Joshi, R.; Tsakalis, K.; McArthur, J. W. and Dash, S. (2014) "Account for Uncertainty with Robust Control Design (I)" *Chemical Engineering Progress*, **11**, pp. 31–38.

Smith, C. (2016) "PID Explained for Process Engineers: Part I: The Basic Control Equation" *Chemical Engineering Progress*, **1**, pp. 37–44.

Smith, C. (2016) "PID Explained for Process Engineers: Part II: Tuning Coefficients" *Chemical Engineering Progress*, **2**, pp. 27–33.

Smith, C. (2016) "PID Explained for Process Engineers: Part III: Features and Options" *Chemical Engineering Progress,* **3,** pp. 51–58.
https://cse.google.com/cse?cx=partner-pub-31769960209562223:6582549258&ie=UTF-8&q=valves&sa=Search& ref=&gfe_rd=ssl&ei=vEA7V4z5Ltek-wXQqYKQBQ#gsc.tab=0& gsc.q=valves&gsc.page=1 (accessed August 29, 2016).

19

Beer Brewing Revisited

Now that we have reviewed most of the basic topics in chemical engineering, let us go back and review the beer flow sheet that we discussed earlier (see Figure 19.1).

First, we have numerous raw materials that enter this process, and the unique aspect to most of them is that they are "natural" in the classical sense. This means that their composition, in the broadest sense of the word, will vary with time, weather, and how they are grown and harvested. But on the other end of this process is desired a constantly unvarying output of a beer whose taste and quality is expected to be the same by its loyal drinkers. This means that the process control embedded in each of the process unit operations must be able to respond to a constantly changing raw material input. Quite a challenge!

What's involved in accomplishing this from a chemical engineering standpoint? Since we know that the raw materials will be constantly changing, we need some way to measure the compositions and adjust the downstream unit operations so that the beer flavor is unchanged. What would we measure? The chemical composition to the extent possible. Particle size and particle size distribution may be important. Bulk and true density would affect the recipe and the size of the batch and percentage volume used in the various tank "reactors." The particle size distribution, both due to the incoming materials and how this is affected by initial grinding and other processing, will affect the reaction rate, so the brew master will make adjustments as necessary. The brew master may not know the fundamentals of reaction kinetics, but it is understood based on experience how these changes will affect the recipe, and the process control system will be adjusted accordingly.

And what is adjusted? What process variables and analysis can link the composition of the process stream as it goes through the process, to the final taste/composition? All the intermediate steps need to be adjusted as well. What temperature, pressure, and vessel reaction times and compositions need to be adjusted? Do changes in the intermediate compositions change viscosity

Chemical Engineering for Non-Chemical Engineers, First Edition. Jack Hipple.
© 2017 American Institute of Chemical Engineers, Inc. Published 2017 by John Wiley & Sons, Inc.

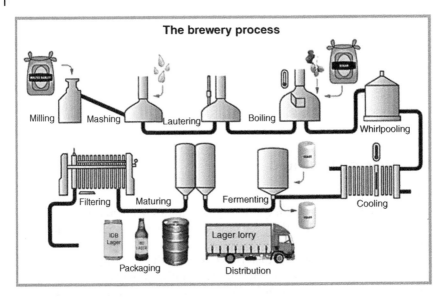

The brewery process

Milling Mashing Lautering Boiling Whirlpooling

Filtering Maturing Fermenting Cooling

IDB Lager Packaging Lager lorry Distribution

Figure 19.1 The brewing process. Source: https://chem409.wikispaces.com/ brewing+process. Used under http://www.creativecommons.org/licenses/by-sa/3.0. © Wikipedia.

and density? If so, how do these changes affect the performance of the heat exchangers? The agitation systems?

Fermentation is a chemical reaction that produces ethanol (C_2H_5OH). Is the reaction rate versus particle size and composition known or only in the head of the brew master? Is the reaction rate as a function of temperature known? The activation energy? How is the residence time in the various downstream vessels determined?

We see heat exchangers in use to drop the temperature after the primary brewing step. How would we decide what kind of heat exchangers? Which fluid would be put on the shell side? Tube side? How do the fluid properties change as a function of the brewing recipe and raw materials? What might foul the exchanger? How would we clean it? The fact that this is a product consumed by people affects this decision. What happens when (not if!) the exchanger leaks? Water into ethanol? Ethanol into water? Diluted beer might be less of a problem than ethanol (which has a biological oxygen demand if released into a public waterway). We at least need to think about it.

Filtering of the process stream is required. What affects this rate and the type of filter media chosen? Should constant pressure or rate be the preferred process? Is vacuum filtration a logical option? Why or why not? What is done with the filter residue? Any environmental issues in this decision? Does it have use as a by-product or agricultural feed? What kinds of filters have been considered? Should this unit operation be revisited? If there is a market for this material, is further processing necessary to change dryness, particle size, or particle size distribution?

Since there is the possibility of biological activity at the end of the process, most beers are pasteurized prior to shipping and inventoried. Destroying residual bacteria is again a chemical process. Many breweries use what is known as "high temperature–short time" pasteurization. If the beer is held at a high temperature or at a more moderate temperature for a long time, the ethanol (its primary ingredient) can also degrade into off-flavor ingredients such as aldehydes and ketones. The kinetic rate curve for this process must be understood so that this last process, necessary to ensure no biological contamination, can be controlled. Coors is a brand name in the beer business. It makes, as a major point of its advertising, that it is "cold brewed." It is also shipped in refrigerated trucks (a significant cost). This type of process does not pasteurize the beer. If kept below a certain temperature, the biological activity is minimized, and if this low temperature is maintained throughout the supply chain, this cold "flavor" can be maintained.

Making beer is a chemical process and has all the same issues of concern, design, and control that make any complicated chemical with a four syllable name.

Appendix I

Future Challenges for Chemical Engineers and Chemical Engineering

This kind of list or discussion will be highly dependent on both the author and their experiences and viewpoints. Having said that, here are some future challenges that chemical will impact with the skills and knowledge of chemical engineers:

1) Energy Resources and Use. In this context, we need to put aside "agendas" that may have political constituent drivers as their prime concern (e.g., corn-based ethanol). Energy is required to feed people, house people, and allow them to move from place to place. Unless we decide to move back to a total agrarian culture and reduce our standard of living, more efficient sources of energy will be required as global population continues to grow and what we refer to now as "Third World" countries demand conventional standards of living long known in the Western world. As a recent example, a breakthrough has occurred fundamentally in understanding of underground fluid flow and mechanics, which we commonly refer to as "fracking," allowing the recovery of hydrocarbons, previously locked in underground rock formations, via the injection of high-pressure fluids to break them apart. Enhanced oil recovery adds additional surface chemistry technology to the basic practice of drilling holes in the ground into reservoirs of liquid.

Energy conservation will be equally important. It is unrealistic to think that Third World populations will not desire the same kind of conveniences that most people in advanced societies enjoy, and if this is to happen, increased energy supplies as well as more efficient energy utilization will be required. We have already seen, in the last few decades, dramatic increases in energy efficiencies across many industries and products, driven in part by the rise in the price of energy. Though this has taken a turn the other way in recent years, once these energy conservation changes are put in place, it is unlikely that they will be reversed. As part of this effort, progress in solar energy conversion will advance steadily. Though solar energy is available globally at a far higher amount than we need, its energy density is miniscule

Chemical Engineering for Non-Chemical Engineers, First Edition. Jack Hipple.
© 2017 American Institute of Chemical Engineers, Inc. Published 2017 by John Wiley & Sons, Inc.

compared to hydrocarbon fuels. There has been steady progress to improve the efficiency of solar power conversion to both thermal and electrical energy, but it is still far from sufficient to support a solar economy. Chemical engineers will play a key role as this technology advances in terms of catalysts, collection material efficiencies, and energy distribution technology.

2) Water. With a growing global population and the small percentage of water resources being immediately usable for drinking water, using non-fresh water drinking supplies as well as recycling and reusing water is a grand challenge. Potable water, in combination with food and energy, is what is sometimes called the "great nexus" of engineering challenges for the 21st century.

3) Materials. There are two long-term aspects to this topic in which chemical engineering will be heavily involved. The first, and most commonly discussed, is plastics recycling. As we discussed in Chapter 17 regarding how polymeric materials are typically made, they are long chains of monomers, typically requiring significant energy input in both their monomer production and the polymer itself. The separation of plastics into easily reusable materials is very costly and inefficient and rarely produces materials that are as usable as the original starting materials. The cost of the separation process itself (polyethylene, polystyrene, ABS, nylon, etc.) is rarely justified economically but is instead subsidized through taxpayer funding. The properties of the polymer, during such a process, typically degrade in some way, primarily through reduction in molecular weight, making them unusable for their original market. We see many examples of products in the store with a label saying "contains up to 10% recycled materials." If more were added, the properties the user is paying for would not be achievable. Several things will happen long term, all requiring significant chemical engineering input. The first is separation technologies that may allow more economical separation of different types of discarded plastics, allowing a higher level of use in original products. The second is a reduction in the amount of plastics used through redesign of packaging and other systems. For example, there may be cases where a large plastic package may be reduced to a "shrink wrap" type of package utilizing far less plastic. It may also be possible to eliminate traditional packaging altogether with various types of coatings. Thirdly, economical pyrolysis processes (heating of materials to high temperature in the *absence* of air, the opposite of combustion) may allow the plastics to be converted back into their starting monomers (i.e., polystyrene to styrene, polyethylene to ethylene, polypropylene to propylene, etc.) producing traditional hydrocarbon monomers that are much easier to separate via traditional chemical engineering unit operations such as distillation and also allow their use in any application, not just for the plastic from which they originated. Work is going on in all of these areas, but the latter is probably the most robust option long term.

Under this heading we must also list "nanomaterials" that are now finding their way into consumer products to provide unique functionality. There is still ongoing debate and data analysis of the potential impact of 10^{-9} particle size materials and their distribution within the environment. However, these size materials have dramatic positive impact on many material properties. The surface chemistry, interface with other materials, and the economical production of such materials will be major challenges for both chemists and chemical engineers. Any separation process will perform significantly different when the particles to be separated are in this size range.

In the periodic table we reviewed earlier, there is a class of compounds known as "rare earth" metals. Examples include metals such as dysprosium and neodymium. Many of these types of materials, in small quantities, are used to improve the properties of base metals. They are rare in their occurrence and expensive to manufacture. There is a major chemical engineering challenge in reducing the cost of their recovery and production, as well as finding alternative materials and product designs that do not require these kinds of rare materials.

4) Medical, Biomedical, and Biochemical Applications. We have all heard of and seen examples of artificial organs now used routinely. If we really consider what artificial hearts and kidney dialysis machines do, they are simply pumps and filters—chemical engineering replacements for original human body parts. All of the concepts we have discussed, including fluid flow, fluid properties, polymers, and filtration, are the basis for the design of these artificial organs. As we progress in understanding of human body functionality, it is likely that we will see additional artificial and replacement organs and body parts where chemical engineering fundamentals are instrumental in their design. Examples could include artificial lungs, replacement joints, and body fluid processing. A core part of any of these types of products are chemical engineering principles related to fluid flow, friction, filtration fundamentals, porosity, pressure drop, and many others. In many schools today, the chemical engineering department has been combined with this type of research and is renamed as biochemical engineering or biomedical engineering.

An artificial heart is a synthetic replication of the normal human heart, but from a chemical engineering viewpoint, it is simply a pump. All the aspects of pump design (friction, energy requirements, flow rates, valve restriction and control, etc.) apply equally to such devices as they do to conventional pumps except that the restrictions on energy available and friction and pressure drop are much more limited. The materials from which we make these devices must also be compatible with human tissue so that they are not rejected. This is a far more sophisticated issue than corrosion concerns discussed earlier as incompatibility is not an issue simply of material degradation but of possible loss of human life.

There are many prescription drugs that are much more desirable to be taken on a small dosage basis continuously rather than large dosages once or twice a day. The encapsulation and slow release of these types of drugs via a skin patch is becoming more common. The encapsulation and slow release of a drug is a result of understanding the mass transfer rates through the skin and how to match this absorption rate with an encapsulation technique that releases the drug at the same rate.

This is in no way a complete list of future chemical engineering challenges and where chemical engineering skills will be useful, and all readers are encouraged to think about their own business and technical world and where chemical engineering principles could be used in a positive way.

Biochemical engineering is the combination of chemical engineering and biology in a way to scale up and produce useful medical products and systems. From a chemical engineering standpoint, there are a number of unique challenges in this task. If we refer back to our review of membrane materials, we see that biological molecules such as viruses and proteins have particle sizes in the range of $0.01–10\,\mu m$, far smaller than the usual particle sizes produced in conventional inorganic and organic chemical reactions. This makes separation and recovery of active, desired molecules far more difficult and requires the use of the far more expensive and sophisticated recovery processes.

Most biochemical entities which are processed start with very dilute solutions. The combination of this factor and the particle size makes liquid–solid separation a major challenge in this area, especially as it relates to pharmaceutical and FDA purity requirements.

Most biological materials exist and perform their function at body and ambient conditions. Thus, it is not possible to run most reactions at higher temperatures where reaction rates might be accelerated. This fact may also put limits on other unit operation concepts for consideration.

The potency of many biopharmaceuticals and drugs is such that the amount of material actually needed for a significant market may be the equivalent of a small pilot plant in the traditional chemical or petrochemical area. Some of the scale-up guidance discussed earlier would not necessarily apply; however, the need and drive for efficient production and high yields of extremely valuable materials is no less important. Challenges in formulating the drug into a suitable form for human consumption, including such variables as its dissolution rate in the stomach and gastric tract, and the inert ingredients required, are substantial.

Stereo specificity is not a concept normally considered in conventional chemical engineering and processing. However, it is very important in the design and manufacture of biologically active molecules and was discussed previously in Chapter 17. The carbon atom, with its normal four bond linkages, is unique in that is, geometrically, a pyramid with the "C" molecule at the center. When the four molecules attached to this carbon are different, there is the possibility that the carbon will be "chiral," meaning that it will not be symmetrical in terms of geometry. One of the side groups will appear (if drawn on a piece of paper) to be going into the paper while the other jutting out. This distinction is known as "left-hand" and "right-hand" rotated molecules (or D-, L-; meaning dextro or levo rotated, respectively). This is important in biologically related or active molecules (such as drugs) in that our human body only recognizes and interacts with L- (levo) rotated molecules. These two different chirally rotated molecules are known as *enantiomers*, meaning that they are not superimposable upon each other, no matter how the molecule is rotated physically.

Our bodies only recognize and process "levo" or "L-" rotated molecules, so everywhere an optical isomer exists, it means that there is an identical molecule produced in the same concentration which may have to be separated prior to the material's use. This "non-absorption" of the opposite isomer is the basis for some artificial sweeteners in the market whose structure is "right handed." We perceive the taste of the molecule (sweetness), but our bodies do not absorb it and it becomes part of normal body fluid waste. The difference between left- and right-handed oriented molecules is described as different "enantiomers."

It is not always necessary to separate the enantiomers (depending on possible negative impact of the other optical isomer), but when it is, there are some important impacts on chemical engineering process decisions. The primary one is that these molecules, though they have different "optical" properties, have the same physical properties such as boiling or melting points. This makes it impossible to separate these enantiomers by such conventional, and lower cost, unit operations such as distillation. There can be small differences in solubility characteristics which can be used in crystallization schemes as well as chromatography and special ion-exchange resins.

In addition to the primary challenges mentioned earlier, we also have the challenge of slow reaction rates. Most biological processes have very slow kinetic rates compared to conventional chemical reactions, and due to the limitations of temperature sensitivity, the ability to raise temperatures during such processes as tissue culture growing is very limited. Agitation is usually required in biochemical reactions as it is in other chemical reactions; however the physical sensitivity to mechanical forces may require special designs.

It is also important to note that chemical engineers, with their strong background in safety, can also contribute to safety analyses of biochemical processes and products. The potential release of harmful, biochemically active materials with small particle sizes, such as viruses, is always of concern, and chemical engineering expertise in gas handling and filtration can be used.

Additional Resources

Benz, G. (2016) "Optimizing Aerobic Fermenter Operation" *Chemical Engineering Progress*, 3, pp. 60–62.

Harrison, R. (2014) "Bioseparation Basics" *Chemical Engineering Progress*, 10, pp. 36–42.

Levesque-Tremblay, G. (2016) "Bio-Molecular Engineering: Little Bio-Machinery to Solve Grand Challenges" *Chemical Engineering Progress*, 4, p. 23.

Ruan, G. and Winter, J. (2012) "Chemical Engineering at the Intersection of Nanotechnology and Biology" *Chemical Engineering Progress*, 12, pp. 38–40.

Tryzbycien, T. and Hooker, N. (2015) "Continuous Processing in Downstream Operations" *Chemical Engineering Progress*, 12, pp. 38–44.

Wibowo, C. and O'Young, L. (2005) "A Hybrid Route to Chirally Pure Products" *Chemical Engineering Progress*, 11, pp. 22–27.

Wispelwey, J. "Drug Delivery and Chemical/Biological Engineering" (2013) *Chemical Engineering Progress*, 3, p. 18.

Appendix II

Additional Downloadable Resources

Note: This list is in no way a comprehensive list, nor does it imply any endorsement on the part of the author or publisher. It does not include purchasable material and books that are available for any of these topics. This list supplements the articles listed at the end of each chapter from AIChE's flagship publication, *Chemical Engineering Progress*. Any industrial web sites listed here are only because of useful general information and illustrations and again do not imply any endorsement by the author or the publisher.

General

http://www.aiche.org
http://www.engineeringtoolbox.com
http://accessengineeringlibrary.com/browse/perrys-chemical-engineers-
   handbook-eighth-edition

1. Chapters 1–3

MSDS sheet requirements: https://www.osha.gov/Publications/OSHA3514.html
MSDS sheet example: propane http://airgas.com/msds/001045.pdf

2. Flammability Limits and Explosion Pressure Information for Compounds

https://en.wikipedia.org/wiki/Flammability_limit
http://www.engineeringtoolbox.com/explosive-concentration-limits-d_423.html
http://www.chemicalbulletin.ro/admin/articole/25469art_13%2858-61%29.pdf
http://www.cdc.gov/niosh/mining/UserFiles/works/pdfs/fompa.pdf

Chemical Engineering for Non-Chemical Engineers, First Edition. Jack Hipple.
© 2017 American Institute of Chemical Engineers, Inc. Published 2017 by John Wiley & Sons, Inc.

3. Accelerating Rate Calorimetry

https://www.youtube.com/watch?v=CoIL_wWx3GQ

4. Periodic Table

http://www.chemicalelements.com/

5. Reaction Kinetics and Equilibrium

https://en.wikipedia.org/wiki/Chemical_kinetics
https://www.khanacademy.org/science/chemistry/chem-kinetics/reaction-
rates/v/rate-of-reaction
http://www.csus.edu/indiv/m/mackj/chem142/kinetics.pdf
https://www.chem.tamu.edu/class/majors/tutorialnotefiles/factors.htm

6. Flow Sheets and Economics

https://en.wikipedia.org/wiki/Process_flow_diagram
https://chemengineering.wikispaces.com/Process+flow+diagrams
http://coade.typepad.com/coadeinsider/2009/06/CEP-May-09-Piping-and-
Instrument-Diagrams-COADE.pdf
http://ocw.mit.edu/courses/chemical-engineering/10-490-integrated-chemical-
engineering-i-fall-2006/projects/eng_econ_lecture.pdf

7. Fluid Flow and Pumps

https://www.khanacademy.org/science/physics/fluids/fluid-dynamics/v/
fluids-part-7
https://en.wikipedia.org/wiki/Fluid_dynamics
https://en.wikipedia.org/wiki/Pump

8. Heat Transfer and Heat Exchangers

https://www.wisc-online.com/learn/natural-science/earth-science/sce304/heat-
transfer-conduction-convection-radiation
http://www.physicsclassroom.com/class/thermalP/Lesson-1/Methods-of-Heat-
Transfer

https://www.khanacademy.org/partner-content/mit-k12/mit-k12-physics/v/
  heat-transfer
https://www.youtube.com/watch?v=iIRbhZY8MpE
https://www.youtube.com/watch?v=Jv5p7o-7Pms
https://www.youtube.com/watch?v=seCA3Awv1Qk

9. Reactive Chemicals

https://www4.uwm.edu/usa/safety/chem/reactive.cfm
http://www.aiche.org/ccps/topics/process-safety-technical-areas/chemical-
  reactivity-hazards/reactive-material-hazards
http://www.ehs.utoronto.ca/resources/whmis/whmis5.htm
https://eta-safety.lbl.gov/sites/all/files/Water%20Chemicals%20-%20
  common%20list.pdf
http://www.usf.edu/administrative-services/environmental-health-safety/
  documents/labsafety-highlyreactive.pdf
https://www.youtube.com/watch?v=sRuz9bzBrtY (Bhopal)
http://people.clarkson.edu/~wwilcox/Design/reac-haz.pdf

10. Distillation

http://www.chem.umass.edu/~samal/269/distill.pdf
https://www.youtube.com/watch?v=gYnGgre83CI
https://www.youtube.com/watch?v=hC1PKRmiEvs
http://www.britannica.com/science/distillation
http://www2.emersonprocess.com/siteadmincenter/PM%20Rosemount%20
  Documents/3051S_ASP_Distillation_Column_Flooding.pdf

11. Other Separation Processes

https://www.cpp.edu/~tknguyen/che313/pdf/chap5-1.pdf
http://www.slideshare.net/abhijitcool18/gas-absorption-53768813
http://www.separationprocesses.com/Absorption/GA_Chp03.htm
https://www.youtube.com/watch?v=jtzoB3MOqxE
http://encyclopedia.che.engin.umich.edu/Pages/SeparationsChemical/Strippers/
  Strippers.html
http://www.academia.edu/5044422/Mass_Balances_on_CO_2_Absorption_
  Stripping_Process_Module_1_Material_Balances_on_Absorption_Stripping_
  Process_Module_Author
https://en.wikipedia.org/wiki/Adsorption

http://www.rpi.edu/dept/chem-eng/Biotech-Environ/Adsorb/adsorb.htm
http://www.lenntech.com/library/adsorption/adsorption.htm
http://www.chem.qmul.ac.uk/surfaces/scc/scat2_3.htm
https://en.wikipedia.org/wiki/Chromatography
http://www.explainthatstuff.com/chromatography.html
http://www.slideshare.net/bejoybj/advanced-chromatography-technique
http://www.novasep.com/technologies/chromatography-for-large-scale-bio-industrial-applications.html
http://www.kochmembrane.com/Learning-Center/Technologies.aspx
http://www.newterra.com/sites/default/files/pictures/newterra_membrane_brochure_sep2014.pdf

12. Evaporation and Crystallization

https://en.wikipedia.org/wiki/Evaporation
http://www.encyclopedia.com/topic/evaporation.aspx
http://www.entropie.com/en/services/evaporation/applications/
https://en.wikipedia.org/wiki/Crystallization
http://www.pharmoutsourcing.com/Featured-Articles/146653-Industrial-Crystallization-of-Pharmaceuticals-Capability-Requirements-to-Support-an-Outsourcing-Paradigm/

13. Filtration

https://en.wikipedia.org/wiki/Filtration
http://www.eaton.eu/Europe/Filtration/index.htm
http://www.quantrol.com/products.php?category=14&subcategory=214
https://www.youtube.com/watch?v=M4wBd1_CvNw
https://www.youtube.com/watch?v=iQAxVqCL2rk
https://en.wikipedia.org/wiki/Rotary_vacuum-drum_filter
http://www.komline.com/docs/rotary_drum_vacuum_filter.html

14. Drying

https://en.wikipedia.org/wiki/Fluidized_bed
https://en.wikipedia.org/wiki/Rotary_dryer
https://en.wikipedia.org/wiki/Rolling_bed_dryer
http://www.bepex.com/systems/thermal-processing/drying/
http://www.gea.com/global/en/binaries/GEA%20Barr-Rosin%20-%20Industrial%20Drying%20Brochure%20-%20English%20-%20US%20letter_tcm11-23426.pdf

15. Solids Handling

http://www.aiche.org/topics/chemical-engineering-practice/solids-handling-particle-technology
http://www.gre.ac.uk/engsci/research/groups/wolfsoncentre/home
http://www.slideshare.net/physics101/storage-bins-and-hoppers
http://www.aiche.org/academy/courses/els102/flow-solids-bins-hoppers-chutes-and-feeders
http://www.inti.gob.ar/cirsoc/pdf/silos/SolidsNotes10HopperDesign.pdf
https://www.aiche.org/sites/default/files/cep/20131125_1.pdf
http://www.chemengonline.com/hopper-design-principles/?printmode=1
http://www.academia.edu/5770816/SCREW_CONVEYOR_BASIC_DESIGN_CALCULATION_CEMA_Conveyor_Equipment_Manufacturer_Association_Approach
http://www.kwsmfg.com/services/screw-conveyor-engineering-guide/horsepower-calculation.htm
http://www.aiche.org/sites/default/files/docs/webinars/JacobK-Pneumatic ConveyingPDFmin.pdf
http://blog.bulk-online.com/general/65.html
http://jenike.com/engineering/pneumatic-conveyors/

16. Tanks and Vessels

http://jenike.com/engineering/pneumatic-conveyors/
https://en.wikipedia.org/wiki/Pressure_vessel
http://www.slideshare.net/ledzung/storage-tanks-basic-training-rev-2
http://www.ec.gc.ca/lcpe-cepa/default.asp?lang=En&n=61B26EE8-1&offset=10&toc=show

17. Polymers and Plastics

https://en.wikipedia.org/wiki/Polymer
http://www.pslc.ws/macrog/kidsmac/basics.htm
https://www2.chemistry.msu.edu/faculty/reusch/virttxtjml/polymers.htm
http://matse1.matse.illinois.edu/polymers/ware.html
https://en.wikipedia.org/wiki/Plastic
https://www.plasticsindustry.org/aboutplastics/
http://www3.weforum.org/docs/WEF_The_New_Plastics_Economy.pdf
http://resource-recycling.com/pru_mag
https://en.wikipedia.org/wiki/Category:Plastics_additives

http://www.intertek.com/polymers/analysis/additives/
http://www.slideshare.net/devraj87india/additives-in-plastics

18. Process Control

https://en.wikipedia.org/wiki/Process_control
http://www.eng.unideb.hu/userdir/deak.krisztian/BOOK.pdf
http://www.itl.nist.gov/div898/handbook/pmc/section1/pmc13.htm
http://www.engineeringtoolbox.com/process-control-systems-t_32.html
http://nptel.ac.in/courses/103105064/
http://www.learncheme.com/screencasts/process-controls

Appendix III

Answers to Chapter Review Questions

Chapter 1 Review: What Is Chemical Engineering?

1   Chemical engineering is a blend of:
  A __Lab work and textbook study of chemicals
  B X Chemistry, math, and mechanical engineering
  C __Chemical reaction mechanisms and equipment reliability
  D __Computers and equipment to make industrial chemicals
  Answer: (b) Chemistry, math, and mechanical engineering. Though chemical engineering includes the other answers, they are only minor parts of a larger description.

2   Major differences between chemistry and chemical engineering include:
  A __Consequences of safety and quality mistakes
  B __Sophistication of process control
  C __Environmental control and documentation
  D __Dealing with impact of external variables
  E X All of the above
  Answer: (e) All of these factors are major differences.

3   A practical issue in large-scale chemical operations not normally seen in shorter-term lab operations is:
  A __Personnel turnover
  B __Personnel protective equipment requirement
  C X Corrosion
  D __Size of offices for engineers versus chemists
  Answer: (c) Corrosion issues are frequently not seen in short term laboratory runs, especially if the lab work is done in glassware. Personnel protective equipment specifications should be the same no matter the scale of the work being done.

Chemical Engineering for Non-Chemical Engineers, First Edition. Jack Hipple.
© 2017 American Institute of Chemical Engineers, Inc. Published 2017 by John Wiley & Sons, Inc.

4 Issues that complicate large-scale daily chemical plant operations to a much greater degree than laboratory operations include all but which of the following

    A __Weather conditions

    B __Emergency shutdown and loss of utilities

    C __Upstream and/or downstream process interactions

    D X Price of company, suppliers, and customer stocks that change minute by minute

Answer: (d) Weather needs to be considered because many large scale chemical process operations are outside without building protection. This subjects process equipment to wide fluctuations in temperature, affecting to some degree any unit operation which is affected by temperature or temperature differential (i.e., reactors, heat exchangers, and distillation columns). Emergency shutdowns and loss of utilities consequences are typically more severe, as well as process changes' impact on directly connected processes. Though prices may be affected by how well we do these things, it is not a direct concern.

5 A chemical engineering unit operation is one *primarily* concerned with

    A __A chemical operation using single-unit binary instructions

    B X Physical changes within a chemical process system

    C __Operations that perform at the same pace

    D __An operation that does one thing at a time

Answer: (b) The concept of "unit operations" generally refers to process operations which involve physical changes (separations). However, in such unit operations as distillation, there are multiple unit operations involved.

Chapter 2 Review: Health and Safety

1 Procedures and protective equipment requirements for handling chemicals include all but:

    A X Expiration date on the shipping label

    B __MSDS sheet information

    C __Flammability and explosivity potential

    D __Information on chemical interactions

Answer: (a) Though an expiration date is a useful piece of information (especially in food and pharmaceutical packaging), it is not a key requirement in this area. The other are far more important.

2 Start-ups and shutdowns are the source of many safety and loss incidents due to:

    A __Time pressures

    B __Unanticipated operational and/or maintenance conditions

    C __Lack of standard procedures for unusual situations

    D X All of the above

Answer: (d) All of these items can and have contributed to significant problems in plant startups.

3 The "fire triangle" describes the necessary elements required to have a fire or explosion. In addition to fuel and oxygen, what is the third item that must be present?

A X Ignition source
B __Lightning
C __Loud noise
D __Shock wave

Answer: (a) An ignition source is required. Shock and lightning may be causes but they are not a general description of what is required. They are sub-sets and examples of ignition sources.

4 The NFPA "diamond," normally attached to shipping containers, indicates all of the following except:

A __Degree of flammability hazard
B __Degree of health hazard
C X Name of chemical in the container
D __Degree of reactivity

Answer: (c) The point of this type of labeling is to give first responders (most of whom will have no knowledge of chemistry) knowledge of the type of material they are dealing with. The name of the chemical is useful, but not important for emergency response with the exception of immediate local responders with whom a local processing facility has excellent close communication.

5 The lower explosive limit (LEL) and upper explosive limit (UEL) tell us:

A X The range of flammability under some conditions
B __The range of flammability under all conditions
C __The upper and lower limits of the company's tolerance for losses
D __The upper and lower limits of the amount of flammable material pumped into a vessel

Answer: (a) UEL and LEL normally refer to limits at atmospheric pressure and the material in pure form. Pressure and the presence of other materials can change the UEL and LEL.

6 Autoignition temperature is the temperature at which:

A __The material loses its temper
B __A material automatically explodes
C X A material, within its explosive range, can ignite without an external ignition source
D __Fire and hazard insurance rates automatically increase

Answer: (c) Autoignition does not mean a material will automatically ignite or explode but it has the potential to do so without deliberate ignition.

7   Toxicology studies tell us all but which of the following:
    A __The difference between acute and long-term exposure effects
    B __Repeated dose toxicity
    C __Areas of most concern for exposure
    D X To what degree they are required and how much they cost
    Answer: (d) These studies give us only data upon which to make decisions.

8   An MSDS sheet tells us:
    A __First aid measures
    B __Physical characteristics
    C __Chemical name and manufacturer or distributor
    D X All of the above
    Answer: (d) As required, an MSDS sheet must include all of this information.

9   A HAZOP review asks all of these types of questions except:
    A __Consequences of operating outside design conditions
    B X What happens to the engineer who makes a bad design assumption
    C __Safety impact of operating above design pressure conditions
    D __Environmental impact of discharge of material not intended
    Answer: (b) Though we might make some assumptions, this is not part of the HAZOP review.

Chapter 3 Review: The Concept of Balances

1   The concept of balances in chemical engineering means that:
    A __Mass is conserved
    B __Energy is conserved
    C __Fluid momentum is conserved
    D X All of the above
    Answer: (d) All aspects must balance.

2   If a mass balance around a tank or vessel does not "close" and *instrumentation readings are accurate*, then a possible cause is:
    A __A reactor or tank is leaking
    B __A valve or pump setting for material leaving the tank is incorrect
    C __A valve or pump setting for material entering the tank is incorrect
    D X Any of the above
    Answer: (d) Any of these are possibilities. If a tank is leaking, there should be some type of alarm system to alert that this is happening. The same should be true of the other possible causes.

3 If an energy balance around a reaction vessel shows more energy being formed or released than should be (*and the instrumentation readings are correct*), a possible cause is:

A __Physical properties of the materials have changed

B _X_ A chemical reaction (and its associated heat effects) is occurring that has not been accounted for

C __Insulation has been added on the night shift when no one was looking

D __A buildup of material is occurring

Answer: (b) Need to check to see if all the possible chemistry is understood.

4 If pressure in a pipeline has suddenly dropped, it may be because:

A __A valve has been shut not allowing fluid to leave

B _X_ A valve has been opened, allowing fluid to leave

C __It has calmed down

D __A downstream process has suddenly decided it would like what is in the pipe

Answer: (b) Note that the gauge itself disappearing does not necessarily mean that the pipe has ruptured.

5 Ensuring accurate measurements of pressure, flow, and mass flows is critical to insure:

A __We know what to charge the customer for the product made that day

B __We know when to order replacement parts

C __We know how to check bills from suppliers

D _X_ Knowledge of unexpected changes in process conditions

Answer: (d) Accurate on line measurements of process flows and conditions can alert us, not only to unexpected changes in process conditions, but to confirm that planned changes have produced the desired results.

Chapter 4 Review: Stoichiometry, Thermodynamics, Kinetics, Equilibrium, and Reaction Engineering

1 Stoichiometry determines ratios and kinetics determine:

A __Kinetic energy

B _X_ Rate

C __Energy release

D __Ratio of rate to energy

Answer: (b) Kinetics and kinetic rate constants determine how fast a particular chemical reaction will proceed.

2 Competitive reactions refer to:

A __Reactions that are also practiced by a competitor

B X Multiple reactions that may occur from the same starting raw materials

C __One or more reactions that compete for raw materials based on price

D __One reaction that runs right after another

Answer: (b) A competitive reaction refers to a situation where more than one reaction, producing different products, can occur.

3 The same raw materials, combined in the same ratio, can produce differing products:

A __Yes

B __No

C __Sometimes, depending upon value of the products produced

D X Yes, depending upon reaction conditions

Answer: (d) The same starting materials, when combined and undergoing a chemical reaction, can produce different products depending upon temperature, pressure, and possibly the presence of a catalyst.

4 Thermodynamics of a chemical reaction determine:

A X The amount of energy released or consumed (needed) if the reaction occurs

B __Under what circumstances a reaction will occur

C __Time delay in a reaction starting

D __How dynamic the reaction is

Answer: (a) The overall thermodynamics of a reaction tell us whether a reaction is exothermic (heat releasing) or endothermic (heat consuming).

5 A kinetic rate constant:

A X Is affected by temperature

B __Is not affected by stoichiometry

C __Is not affected by altitude

D __Is affected by size of reaction equipment

Answer: (a) The kinetic rate constant is a function of temperature; the reaction rate may be affected by the other conditions.

6 The rate of a chemical reaction:

A __Can be changed by changing pressure and/or temperature,

B __Will be affected by stoichiometry and ratios of reactants

C __Will be affected by how fast products are removed

D X All of the above

Answer: (d) Any of these variables can affect the actual rate of a reaction (i.e. mole/h. converted or produced). Anything we do to affect actual stoichiometry during a reaction will have an effect.

7 The rate of a chemical reaction is typically_____with temperature:
   A __Linear
   B __Quadratic
   C X Logarithmic
   D __Semilogarithmic
   Answer: (c) Though there can certainly be specific chemical reaction mechanisms that make a generalization a challenge, most chemical reactions respond to a temperature change logarithmically. This means that the rate of reaction responds dramatically to temperature. This is a major concern with exothermic reactions which generate heat. As their rate increases, the rate of heat release increases in the same way, generating possible reactive chemicals concerns.

8 Conversion of a chemical reaction will always be:
   A X The same or greater than yield of the same reaction
   B __Less than the selectivity to multiple reaction products
   C __Unaffected by the kinetic rate constant
   D __Different from the selectivity of a reaction
   Answer: (a) Conversion refers to the amount of a raw material, entering a chemical reaction, which converts to another product in some percentage. Yield typically refers to conversion to the desired product. Conversion at 100% to desired product would equal 100% yield, but not less.

9 If a calculated heat of a particular reaction is negative (exothermic), it means:
   A __We don't want the reaction to occur
   B __The heat calculation is incorrect as it should be a positive number
   C X Energy is released if the reaction occurs
   D __Energy is required to sustain the reaction
   Answer: (c) An exothermic reaction is one which has a net release of energy as calculated as the difference between the heats of formation of its products minus its reactants.

10 If a calculated heat of a particular reaction is positive (endothermic), it means:
   A __It is good for the reaction to occur
   B X Constant energy input is required to sustain the reaction
   C __The reaction will never stop once started
   D __All of the above
   Answer: (b) Endothermic reactions have a positive difference between the heats of formation of products and reactants, thus requiring a constant input of energy to sustain them.

11   Equilibrium in a chemical reaction system can be affected by:
A __Ratio of reactants
B __Temperature
C __Number of possible reactions
D X All of the above
Answer: (d) The equilibrium in a reaction system will be affected by any of these variables, as they all affect the equilibrium constant of a reaction. This situation becomes even more complicated if there is the possibility of more than one chemical reaction occurring.

12   The equilibrium constant K_e refers to:
A __The ratios of reactants to products
B __The ratio of reactants to products under certain conditions
C __The ratio of products to reactants
D X The ratio of products to reactants under specific conditions
Answer: (d) The equilibrium constant for a given reaction is a ratio of products divided by reactants. This constant will change with temperature, and for gases, pressure as well.

13   A change in pressure will most likely affect reaction equilibrium for:
A __Liquid–liquid reactions
B __Liquid–solid reactions
C X Gas–gas, gas–liquid, or gas–solid reactions
D __Reactions using a gas whose price is increasing
Answer: (c) Gases change volume and solubility with pressure, so any system using gases in a reaction, regardless of the type, will be affected by pressure.

14   The total time for a reaction to go to completion is affected by all of these except:
A __Kinetic rate constants
B __Rate of heat removal in an exothermic reaction
C __Stoichiometry of reactants
D X Size of the reactor
Answer: (d) The reaction time is affected by all except the size of the reactor. The reactor size relates to how much of a given reaction conversion or yield we want to accomplish in a given vessel, but it is a consequence of the time calculation, not a cause of it.

15   Catalysts can do these things:
A __Lower the temperature or severity of conditions of a reaction
B __Initiate an exothermic reaction
C __Favor one product over another in a reaction system
D X All of the above
Answer: (d) Catalysts are capable of doing any or all of the above, thus increasing the capability of a reaction system to increase yield or reduce temperature and pressure of reaction systems.

16 The loss of catalyst effectiveness over time is most likely due to:

A __Change in stoichiometry in the feed

B X Poisoning or contamination

C __Change in catalyst vendors

D __The introduction of arsenic into the feed

Answer: (b) Though any of these items listed are possibilities, when a long term steady loss in catalyst efficiency is most likely due to some form of contamination or poisoning that needs to be identified and/ or removed.

Chapter 5 Review: Flow Sheets, Diagrams, and Materials of Construction

1 The level of detail contained in a flow sheet, in order of increasing complexity, is:

A __P&ID, mass and energy balance, 3D

B __Mass and energy balance, P&ID, 3D

C X Block flow, mass and energy balance, P&ID, 3D

D __3D, P&ID, mass and energy balance, block flow

Answer: (c) A block flow diagram simply shows connections between process units. Then a mass and energy balance adds numbers to this. A P&ID adds control and instrumentation design, and finally a 3D diagram shows all of this in three dimensions, allowing a much better understanding of the practical aspects of accessing, maintaining, and running the process.

2 Process flow diagrams are important because they:

A __Ensure disk space is used on a process control computer

B X Provide a sense of process stream and equipment interactions

C __Provide a training exercise for new engineers and operators

D __Make effective use of flow sheet software

Answer: (b) Though the other answers are true to some degree, the key value is in showing clearly the interactions between various pieces of equipment, their relationship to each other, as well as a framework to discuss interactions.

3 3D process diagrams are most important because they:

A X Enable personnel to envision the interaction between people and equipment

B __Allow the use of 3D glasses from the movies that otherwise would be thrown away

C __Enable the use of 3D software

D __Show the best location for a security camera

Answer: (a) A three dimension visualization of process equipment can provide valuable information on the practical aspects of how operating personnel can reach valves, escape under emergency conditions, and optimize maintenance.

4 It is important to ensure flow sheets are up to date because:

A __They are used by maintenance personnel to identify connections and equipment

B __They show safety valves and relief systems

C __They are a means of common communication between engineers, operators, and maintenance personnel

D X All of the above

Answer: (d) All of these reasons are important! Frequently changes are made to a process during maintenance or shut downs and the flow sheets are not updated. This is a potential cause of safety and operational hazards.

5 Accurate measurement and knowledge of corrosion rates, as well as what affects them, within process equipment is important because:

A __Pipe vendors need to know when to schedule the next sales call

B __Corrosion meters need to be tested once in a while

C X It is important to understand the estimated life of process equipment and the potential for corrosion products to contaminate process streams

D __We need to keep evacuation plans up to date for equipment failures

Answer: (c) Equipment lifetime and corrosion products entering process streams are both critical concerns.

6 A process fluid with higher water content than one with a lower water content:

A __Will be more corrosive

B __Will be less corrosive

C __Depends on the temperature

D X Can't tell without laboratory data

Answer: (d) Though there are many examples where instincts on corrosion are valid, there are just as many where it is not supported by data. This is what we need and under the conditions the process will run. Example: dry chlorine gas can be handled in steel, but it will "ignite" within a titanium pipe due to a chemical reaction. If the chlorine is wet, steel pipe will rapidly corrode, while titanium has a long life.

Chapter 6 Review: Economics and Chemical Engineering

1 The cost of manufacturing a chemical includes:

A __Capital cost (cost of building the plant)

B __Cost of raw materials

C __Taxes, labor, supplies

D X All of the above

Answer: (d) All of these factors will affect the cost of manufacture.

2   The most important factors in determining the variable cost of manufacture are typically:
    A __Shipping costs
    B __Labor contract changes
    C X Raw material and energy costs
    D __Security
    Answer: (c) The other factors are usually not as significant as raw materials and energy. There may be very special exceptions when shipping costs and siting of a plant to take advantage of this might be a major factor.

3   If capital costs are 50% of the total cost of manufacture, and the production rate is reduced by 50%, the impact on the product's cost of manufacture will be:
    A __10%
    B X 25%
    C __50%
    D __75%
    Answer: (b) $0.5 \times 0.5 = 0.25$ or 25%.

4   If the cost of one raw material, representing 20% of a product's total cost, is raised by 25%, the impact on total cost will be:
    A X 2.4%
    B __4.4%
    C __5.4%
    D __6.4%
    Answer: (a) If this raw material represents 20% of the total cost, then 80% is represented by other factors. If this 20% (0.2) is raised by 25%, we now have 0.2×0.25 or a 5% (0.05). Assuming nothing else changes, the total cost is now 1.05 and this cost impact will be 0.25/1.05 or 2.4%.

Chapter 7 Review: Fluid Flow, Pumps, and Liquid Handling and Gas Handling

1   Total fluid pressure is measured by the sum of:
    A X Static pressure and dynamic pressure
    B __Dynamic pressure and fluid density
    C __Static pressure plus anticipated friction loss
    D __All of the above
    Answer: (a) Static pressure (height) plus dynamic pressure (due to flow pressure).

2 Laminar flow implies:
   A __Pressure drop for the fluid flow is high
   B __Fluid is wandering around with no direction
   C __Pipes are made from plastic laminates
   D X Little or no mixing across the cross-sectional area of a pipe
   Answer: (d) The flow rate is low enough so as to prevent any significant mixing across the diameter of the pipe.

3 Turbulent flow implies:
   A __Fluid is well mixed across the cross-sectional area of the pipe
   B __Pressure drop will be higher than laminar flow
   C __There is little or no adhesion between the fluid and the piping wall
   D X All of the above
   Answer: (d) All of these conclusions are valid in turbulent flow.

4 Turbulent versus laminar flow will affect all but:
   A __Pressure drop in the pipeline
   B __Mixing in the pipe
   C X Cost of the piping materials
   D __Pressure drop across valves and instrumentation
   Answer: (c) Though turbulent flow could imply higher pressure, this not necessarily the case, so the cost of the piping materials themselves are not necessarily affected.

5 Key fluid properties affecting fluid handling systems include all except:
   A __Density
   B __Viscosity
   C X Residence time in tank prior to pumping
   D __Temperature and vapor pressure
   Answer: (c) The residence time in the tank will not affect, but the height of the liquid in the tank could.

6 Viscosity characterizes a fluid's resistance to:
   A __Being pumped
   B __Being held in storage
   C __Price change
   D X Shear
   Answer: (d) The basic definition of viscosity.

7 The viscosity of a fluid is most affected by:
   A __Density
   B __Pressure
   C __Index of refraction
   D X Temperature
   Answer: (d) Temperature has a logarithmic effect; the other variables have little or no effect.

8 The viscosity of an ideal (Newtonian) fluid reacts to a change in shear at constant temperature by:

A X Remaining the same

B __Increasing

C __Decreasing

D __Need more information to answer

Answer: (a) An ideal fluid does not change viscosity with shear.

9 A dilatant fluid's viscosity _____ with shear:

A X Increases

B __Decreases

C __Stays the same

D __Depends on what kind of shear

Answer: (a) Increases (meaning it will thicken).

10 A thixotropic fluid responds to shear by _____ its viscosity:

A __Increasing

B X Decreasing

C __Not affecting

D __Depending on what kind of shear

Answer: (a) Decreasing.

11 In general, adding solids to a liquid (converting it into a slurry) will _____ its viscosity:

A X Increase

B __Decrease

C __Not affect

D __Can't be known

Answer: (a) Increase (in general). There can be situations, depending upon particle size, concentration, liquid surface tension, and nature of the fluid where this may not be the case, but this would be the starting assumption.

12 The Reynolds number is:

A __Dimensionless

B __A measure of turbulence in flow

C __The ratio of diameter × density × velocity

D X All of the above

Answer: (d) All of the above.

13 A dimensionless number in chemical engineering:

A __Provides a simple way of characterizing an aspect of design

B __Has no units (if calculated correctly)

C __Allows a chemical engineer to estimate relative behavior of an engineering system

D X All of the above

Answer: (d) All of the above.

14 Friction in fluid flow is influenced by all but:
A __Fluid properties
B __Flow rate
C __Piping design characteristics
D X Cost of energy to pump the fluid
Answer: (d) Cost of energy does not affect fluid properties (other than their cost!).

15 Pressure drop in a fluid system can be affected by:
A __Length of piping
B __Number and nature of connections and valves
C __Degree of corrosion on walls
D X All of the above
Answer: (d) The longer the pipe, the greater the pressure drop; each valve or connection adds pressure drop; corrosion can decrease pipe diameter over time, increasing pressure drop.

16 The difference between a centrifugal and positive displacement pumps is:
A __Centrifugal pumps are less expensive
B __Positive displacement pumps have a characteristic curve
C X Centrifugal pumps generate constant pressure; positive displacement pumps put out constant flow
D __It is harder to "dead head" a positive displacement pump versus a centrifugal pump
Answer: (c) Centrifugal pumps balance flow and pressure; positive displacement pumps displace a given amount of volume under constant pressure.

17 Centrifugal pumps require a minimum net positive suction head (NPSH) to operate; otherwise they will cavitate. This can be caused by all but:
A __Improper placement of the pump on an engineering drawing
B X Reducing the level of the liquid feeding the pump
C __Raising the level of a tank into which the pump is discharging
D __Raising the temperature of the feed liquid and raising its vapor pressure
Answer: (b) Reducing the level of the feed will decrease the NPSH available.

18 If the process needs to exceed the minimum NPSH available, what options are available?
A __Raise the height of inlet stream to the pump
B __Lower the temperature of the inlet feed
C __Increase the size of the piping in the system
D X Any of the above
Answer: (d) Any of these suggestions will increase the NPSH available, but more than one may be required.

19 The choice of a flow meter will depend upon:
    A __Accuracy required
    B __Pressure drop tolerance
    C __Cleanliness of fluid
    D X All of the above
    Answer: (d) All of these factors need to be considered in choosing a flow meter. Dirty fluids will rule out some, high ΔP will rule out others, and accuracy of measurement needed will also be a screening mechanism. Corrosion resistance will also be a factor.

Chapter 8 Review: Heat Transfer and Heat Exchangers

1 An energy balance around a process or piece of equipment requires knowledge of all but:
    A __Flow rates and temperatures of flows in and out
    B X The speed of the pump feeding the vessel
    C __Heat generated by any reaction occurring
    D __Heat capacities of streams in and out
    Answer: (b) The speed of the pump may contribute to the heat balance, but it will be reflected in the data in (a).

2 The three methods of heat transfer are all but:
    A __Conductive
    B __Convective
    C X Convoluted
    D __Radiation
    Answer: (c) Maybe in thinking, but not in heat transfer.

3 Conductive heat transfer refers to heat moving:
    A __Above
    B __Below
    C __Around
    D X Through
    Answer: (d) All the others would be part of convective heat transfer.

4 Convective heat transfer refers to heat moving:
    A X In a bulk fashion
    B __Only as a function of convective currents
    C __On its own
    D __None of the above
    Answer: (a) Refers to heat transfer occurring through bulk mixing of fluids or gases.

5 Variables that affect the rate of heat transfer are:
 A __Flow rates
 B __Physical properties of fluids
 C __Turbulence within the heat transfer area or volume
 D X All of the above
 Answer: (d) All of these will affect.

6 If the pipe diameter is increased and all other variables remain the same, the rate of heat transfer:
 A __Will increase
 B X Will decrease
 C __Will stay the same
 D __Can't tell without more information
 Answer: (b) If everything else stays the same, the Reynolds number and turbulence will decrease, reducing the heat transfer coefficient.

7 If the viscosity of fluids on the shell side is increased, the heat transfer rate:
 A __Will increase
 B X Will decrease
 C __Will stay the same
 D __Can't tell without more information
 Answer: (b) Any increase in viscosity of fluids or gases anywhere within the heat exchanger will decrease the heat transfer coefficient.

8 The utility fluid is:
 A __The fluid that costs less since it is asked to do anything
 B __A fluid that can move in either direction
 C X The non-process fluid in a heat exchanger
 D __A utility that is a fluid
 Answer: (c) Refers to the cooling or heating fluid that is not the process fluid.

9 The overall heat transfer coefficient includes the resistance to heat transfer through:
 A __The pipe wall
 B __The barrier layer on the shell side
 C __The barrier layer on the tube side
 D X All of the above
 Answer: (d) The overall coefficient includes all of these. It is possible to measure the individual components if desired.

10 Design issues with heat exchangers include all but:
  A __Area required
  B __Corrosion resistance to the fluids
  C __Leakage possibilities
  D X The dollar to euro conversion at the time of design
  Answer: (d) Only ChE's in the financial world impact this; all the other issues are part of design considerations.

11 The primary design limitation of air cooled heat exchangers is:
  A __Fan speed
  B __Distance from a river or lake
  C X Temperature of outside air
  D __Contractor's ability to raise or lower the heat exchanger
  Answer: (c) Although the other factors can have an impact, the primary limitation is the temperature of the ambient air.

12 Fouling and scaling on a heat exchanger can be caused by:
  A X Deposition of hard water salts
  B __Softness of the heat exchanger material
  C __Use of distilled water as a coolant
  D __Poor maintenance
  Answer: (a) Inverse solubility of hard water salts are a primary cause.

13 Radiative heat transfer can be an important concern in:
  A __Sunburns while working in a chemical plant in Houston
  B __Chemicals that are red or yellow
  C __Insufficient heat transfer on a cloudy day
  D X High temperature processing in the oil and petrochemical industry
  Answer: (d) Unless the process temperatures are in the neighborhood of 1000°C, it is unlikely that radiant heat transfer will be a significant issue. Radiation heat transfer is proportional to the fourth power of temperature.

14 High temperature heat transfer fluids are used when:
  A __Cold ones are not available
  B X It is necessary to transfer heat at high temperature and low pressure
  C __Hot water is not available
  D __The plant manager owns stock in a company that makes and sells them
  Answer: (b) Transferring heat under these conditions would ordinarily require high pressure steam, which is costly.

15 The downside of high temperature heat transfer fluids include all but:

A __Flammability

B __Possible degradation and need to recharge

C __Potential chemical exposure to the process fluid

D X Ability to transfer high temperature heat at low pressure

Answer: (d) This is their primary positive.

Chapter 9 Review: Reactive Chemicals Concepts

1 Reactive chemicals reviews start with an understanding of:

A __How reactive management is to safety incidents

B __A summary of last quarter's reactive chemicals incident reviews

C X The chemical stability of all chemicals being handled

D __The cost of changing storage conditions for gas cylinders

Answer: (c) A basic understanding of the chemical, thermal, and physical stability of all materials being handled is a mandatory first step in this process.

2 Reactive chemicals analysis would include all but the following:

A X Management's reaction to a reactive chemicals incident

B __Shock sensitivity

C __Temperature sensitivity

D __Heat generation during any processing

Answer: (a) Management's reaction may be relevant in terms of what is actually done, but it is not a part of the review process.

3 When considering the reactive chemicals potential of an exothermic chemical reaction, the key consideration is:

A __The cost of cooling vs. heating

B __The cost of relief devices and environmental permits relating to an over-pressured reactor

C X The rate of heat generation vs. the rate of cooling required

D __The possible rise in cost of processing

Answer: (c) A reactive chemical incident is caused by the rate of heat generation being faster than the system's ability to absorb and/or remove the heat, so understanding these relative factors is critical.

4 The reason there is a basic conflict between kinetics and heat transfer is that heat transfer is a linear function and kinetics or reaction rates are typically:

A X Logarithmic with temperature

B __Inversely proportional to pressure

C __Subject to residence times in the reactor

D __Vary with the square root of the feed ratios

Answer: (a) Reaction rates normally increase logarithmically with temperature, meaning that the potential for heat generation is greater than the heat removal rate increase with temperature, which will be linear.

5 A rise in reactor temperature will:
   A __Increase the rate of heat removal from the reactor
   B __Increase the rate of any chemical reaction occurring
   C __Lower the viscosity of any liquids in the reactor, increasing the heat transfer rate
   D X All of the above
   Answer: (d) A temperature rise will do all of the above. The drop in viscosity (discussed in the chapter on fluid flow) will also potentially increase the heat transfer coefficient, so there may be a slight increase in heat removal rate.

6 An increase in volume used within a chemical reactor will have what effect on the potential for a reactive chemical incident?
   A __None
   B __Make the system less susceptible
   C X Make the system more susceptible
   D __Need additional information to answer
   Answer: (c) An increase in volume used could increase the heat transfer area available for heat removal (depending on the design of the heat transfer system), but it will also greatly increase the potential energy release from a runaway reaction.

7 A drop in temperature within the reactor will _____the probability of a runaway reaction.
   A __Increase
   B X Decrease
   C __Make no difference
   D __Need additional information
   Answer: (b) Any drop in temperature will reduce the rate of reaction. It will also reduce the rate of heat transfer removal, but remember the first is logarithmic and the second linear.

8 Improper storage of materials in warehouses can be a source of reactive chemicals incidents if:
   A __Moisture-sensitive materials are stored under a leaky roof
   B __Oxidizers and reducers are stored next to each other
   C __Known compound stability time limits are exceeded
   D X Any or all of the above
   Answer: (d) Any of these conditions, or their combination, could create a reactive chemicals concern.

Chapter 10 Review: Distillation

1   Distillation is a unit operation based on differences in:
    A __Solubility
    B __Density
    C X Vapor pressure
    D __Crystallinity
    **Answer: (c) Vapor pressure. The difference in volatility is what we use
    in distillation. The other properties listed can be used in other separa-
    tion processes, especially liquid–solid systems.**

2   A material, solution, or mixture boils when:
    A __The solution is rolling around and bubbling violently
    B X The sum of all the partial pressures equals the total pressure
    C __It's mad
    D __The partial pressures exceed the external pressure by 10%
    **Answer: (b) When the sum of all the partial pressures equals the
    system pressure, the solution boils. It is important to remember that
    the system pressure may not be atmospheric pressure.**

3   The key determinant in how easy it is to separate a mixture by distillation
    is the:
    A __Volatility of the relatives of the mixture
    B __Whether the relatives want to be separated
    C __Ability to heat selectively the most volatile component
    D X Relative volatility of its various components
    **Answer: (d) The volatility of one liquid component compared to others
    in the system is what determines the ease of separation by distillation.
    The higher the volatility of one component is to another (i.e., higher
    relative volatility), the easier it is to separate the components by
    distillation.**

4   On a graphical plot of a distillation system, the 45° line represents:
    A X The vapor phase and liquid phase having the same composition
    B __All phases are created equal
    C __One component has 45% more volatility than another
    D __One component has 45% less volatility than another
    **Answer: (a) This is the line at 45° from the horizontal and represents
    where the liquid and gas have the same composition. In distillation, it
    is useful in analyzing total reflux situations and minimum number of
    stages required to perform a particular separation.**

5 In a two-component distillation system where the relative volatility is displayed on a y–x graph, a higher relative volatility will be displayed, vs. a 45° line, as:

A __No difference in the lines

B __A small difference in the lines

C X A large difference between the lines

D __Price of company, suppliers, and customer stocks that change minute by minute

Answer: (c) A large difference displays a large relative volatility; a small difference a small one. Smaller relative volatilities indicate more difficult distillation separations.

6 In a batch distillation system, the maximum number of stages of separation possible is:

A X One

B __Depends on relative volatility

C __A function of heating rate

D __A function of the batch size

Answer: (a) Since there is only one stage of vaporization, there is only one state of separation.

7 In a conventional continuous distillation system, the top of the column will always contain:

A __A higher concentration of the less volatile component

B X A higher concentration of the more volatile component

C __A higher concentration of the less dense material

D __A higher concentration of the material desired by the customer

Answer: (b) Since the more volatile component will concentrate in the vapor phase, it will be most concentrated at the top of the column.

8 Returning reflux to a distillation column allows:

A __More energy to be wasted

B __More cooling water to be wasted

C X Multiple vaporizations and condensations, yielding purer top and bottom products

D __More capital expenditures to be wasted on a reboiler and condenser

Answer: (c) Condensing some of the overhead product provides the condensing mechanism for the "boil/condense" aspect of distillation. Without reflux, we have no distillation beyond one stage of separation.

9 Increasing reflux to a distillation column results in:

A __Higher pressure drop

B __Purer overhead product

C __More cooling water to be used

D X All of the above

Answer: (d) Increasing reflux does all of these things. The economics and equipment limitations will determine the degree to which increasing reflux can be increased and is desirable.

10 Decreasing reflux to a distillation column results in all except:

 A __Less cooling water and reboiler steam use
 B __Lower overhead purity
 C X Less intensive process control
 D __Pressure drop across the column decreases

 Answer: (c) Decreasing reflux results in all of these items except process control. There is no reason to expect that the process control will be any less demanding.

11 The "operating line" of a distillation column represents a graphical display of:

 A __The line drawn by the process operators when the process computers are offline
 B X The mass balance within the column
 C __The line of code that operates the column
 D __The line that no one on the operating floor is allowed to cross

 Answer: (b) The operating line is a plot of the liquid and vapor composition within the column as opposed to a vapor pressure curve. The distance between it and the vapor–liquid equilibrium line is an indirect indication of the ease of separation by distillation.

12 Varying the reflux ratio in a distillation column allows us to:

 A __Adjust the quality of overhead and bottom products
 B __React to changes in feed compositions
 C __Allow process adjustments to upstream and downstream processes
 D X All of the above

 Answer: (d) Varying reflux ratio allows all of these.

13 Vacuum distillation can result in all but:

 A __Increased energy use
 B __Separation of azeotropes
 C __Separation of high boiling components
 D X Smaller distillation columns

 Answer: (d) The use of vacuum can be useful in all of the areas, but the consequence of using vacuum is larger diameter columns, not smaller. The gas law studied earlier predicts that as pressure is reduced, volume is increased.

14 Azeotropes are:

 A __Special mixtures of chemicals that come from the tropics
 B __Mixtures of chemicals with close boiling points
 C X Mixtures of materials whose vapor composition when boiled is the same as the starting liquid
 D __Impossible to separate

 Answer: (c) When the liquid composition and vapor composition at boiling is the same, we have what is known as an azeotrope. Such a limitation can limit the effectiveness and options in the use of distillation.

15 Ways of separating azeotropes include:
  A __Changing pressure
  B __Using an alternative separation technique
  C __Adding a third component that shifts the vapor–liquid equilibrium
  D X All of the above
  Answer: (d) Any of these techniques can possibly be used to "break" an azeotrope. Which one is chosen is a function of utility costs, the cost of an alternative separation unit operation, and the limitations of adding additional materials to the system.

16 Bubble cap trays in distillation columns have this key advantage:
  A __They trap bubbles
  B X Prevent liquid from dropping down on to a lower tray without contacting vapor
  C __Relatively expensive
  D __High pressure drop
  Answer: (b) Because the bubble cap floats up and down against the incoming vapor stream from the tray below, it prevents liquid on the tray from "leaking" down on to the lower tray.

17 Sieve trays have this disadvantage:
  A __Low pressure drop
  B __Inexpensive and easy to fabricate
  C X Can allow weeping and mixing between stages
  D __Can allow low molecular weight materials to leak through
  Answer: (c) Though the other factors are positives, the fact that there is positive mechanical seal against the down flowing liquid allows some liquid to drop down to the tray below, causing some equilibration and loss of efficiency. Hydraulic design of these type of trays is important.

18 Loose packings used in place of trays:
  A __Will usually have lower pressure drop
  B __Can be more corrosion resistant
  C __Are more likely to breakup due to mechanical shock
  D X All of the above
  Answer: (d) All of these items are characteristic of packed towers vs. tray towers. Metal packings are less susceptible to breakage during installation and handling than ceramic packings.

Chapter 11 Review: Other Separation Processes

1   Absorption is the process for recovering a gas into a:
   A __Solid
   B __Another gas
   C **X** Liquid
   D __Any of the above
   Answer: (c) Liquid. Other unit operations are used in the other choices.

2   Stripping is removal of a gas from:
   A **X** Liquid
   B __A reactor
   C __A tank truck
   D __Any of the above
   Answer: (a) This is the opposite of absorption.

3   The key variable that is used in designing an absorber or a stripper is the:
   A __Temperature of the liquid
   B __Temperature of the gas
   C **X** Henry's law constant
   D __External temperature
   Answer: (c) Henry's Law Constant. This is the ratio of vapor pressure to mole fraction (in the liquid) at any given temperature.

4   Henry's law constant represents:
   A **X** The ratio of gas partial pressure to gas concentration dissolved in the liquid
   B __The inverse of Henry's law variable
   C __The approval of Henry to the gas solubility data generated in the lab
   D __How much more gas will dissolve in a liquid if the pressure is increased
   Answer: (a): This ratio is greatly affected by temperature. The relative Henry's Law Constants tell us to what degree one gas will preferentially be absorbed (or stripped) vs. other gases.

5   In an absorber, the gas enters at the:
   A __End
   B __Top
   C **X** Bottom
   D __Middle
   Answer: (c) Gas, with a contaminant or material for recovery, enters at the bottom and is scrubbed by the liquid entering at the top.

6 In a stripper, the liquid enters at the:
   A __End
   B X Top
   C __Bottom
   D __Middle
   Answer: (b) The liquid, containing a material that cannot be discharged or is recoverable, is fed into the top and gas entering the bottom strips or removes this material.

7 In designing an absorber it is important to take into account:
   A __Proper distribution of inlet gas across the bottom of the tower
   B __Proper distribution of liquid over the cross sectional area of the tower
   C __Potential temperature rise due to heat of gas dissolution
   D X All of the above
   Answer: (d) All of these variables are important to ensure optimum contact between liquid and gas, as well as to take into account changes in Henry's Law constant due to heat of solution of the gas into the liquid.

8 Demisters may be required at the top of a stripper due to:
   A __Fill in void space
   B X Control of possible liquid carryover
   C __Operators are sad when seeing material being removed from a liquid
   D __To supply pressure drop
   Answer: (b) Demisters are fine wire devices designed to coalesce fine particle sized mists and collect and send them back to the column.

9 Adsorption is the process for recovering a component from a fluid or gas onto a:
   A X Solid
   B __Membrane
   C __Liquid
   D __Any of the above
   Answer: (a) Adsorption is the recovery of a component in a gas stream on to a solid adsorbent. Gas recovery into a liquid is absorption.

10 The efficiency of adsorption is governed by:
   A __What kind of carbon is used
   B __Affinity of the gas for the solid
   C __Pore size of the adsorbent
   D X All of the above
   Answer: (d) All of these variables are important and are most often measure in conjunction with an adsorbent supplier.

11 Variables that can affect the efficiency and selectivity of adsorption include:

A __Temperature

B __Pressure

C __Adsorption isotherms

D X_All of the above

Answer: (d) All of these variables are important and can be used individually or collectively to design or improve an adsorption process.

12 Adsorption beds can be regenerated by all of these techniques except:

A __Change in pressure

B __Change in temperature

C X_Seriously wishing

D __Purging with a large amount of gas to displace the adsorbed material

Answer: (c) If only it were that easy! Any of these techniques, individually or in combination can be used to design or optimize bed regeneration.

13 Liquid chromatography is a unit operation that utilizes_____to recover and/or separate liquid components:

A __Molecular size

B __Surface charge

C __Liquid–solid surface chemistry

D X_Any of the above

Answer: (d) Any of these properties can be used.

14 Ion exchange processes use what functionality bound to a polymer surface to achieve separation:

A X_Ionic charge

B __Pore size

C __Differing molecular weight polymer additions

D __Surface roughness

Answer: (a) Ion exchange uses differences in charge and polarity to recover opposite charged materials.

15 Ion exchange beds are regenerated through the use of:

A __Change in pressure

B __Change in temperature

C X_Large volumes of the opposite original charge solutions

D __Purging with a large amount of gas to displace the exchanged material

Answer: (c) This operation replaces the original charge on the resin.

16 A serious practical issue when regenerating ion exchange beds is:
 A __Using the wrong regenerant solution
 B X Hydraulic expansion
 C __Noise created
 D __Regenerating the wrong bed
Answer: (b) Ion exchange resins are chemically cross linked, not allowing them to dissolve; instead most resins have the capability to absorb liquids and thus can hydraulically expand, especially when regenerated. Extra volume must be allowed to accommodate this physical property.

17 Liquid–liquid extraction is a unit operation involving the use of a material's preference to be dissolved in:
 A __One liquid close to its boiling point versus another liquid at room temperature
 B __One liquid close to its freezing point versus another liquid at room temperature
 C __One liquid close to its critical point versus another liquid near its boiling point
 D X One liquid versus another liquid
Answer: (d) While temperature does play a role in design, the primary physical property variable is the relative solubility of one liquid vs. another in a third liquid.

18 To design a liquid–liquid extraction process, the following is needed:
 A __A ternary phase diagram
 B __Knowledge of densities and density differences
 C __Surface tension of process fluids
 D X All of the above
Answer: (d) Knowledge of the mutual solubilities, density differences (which liquid is fed at the top for a continuous column) and surface tension (affecting the need for surfactants) and time needed to separate phases after contacting are all important.

19 Operating and design variables for a liquid-liquid extraction operation include:
 A __Temperature
 B __Contact time
 C __Liquid physical properties
 D X All of the above
Answer: (d) In addition to the physical property data, the operational variable of temperature (and variation), contact time, and physical properties are all needed to design a practical liquid-liquid extraction unit.

20 Leaching is a unit operation used to:
    A __Go back to the days of the gold rush
    B __Recover money from a stingy relative
    C **X** Recover a material from a solid via liquid contact
    D __Recover a material from a solid via gas contact
    Answer: (c) Leaching is used to recover or remove a material from a solid by contacting it with a solution which has a preferential solubility for a component of the solid to be removed or recovered.

21 Membranes separate materials based on the difference in their:
    A **X** Molecular weight and size
    B __Desire to go through a very small hole
    C __Value and price
    D __Cost
    Answer: (a) Membranes, having different pore sizes, are used to separate different size gas molecules.

Chapter 12 Review: Evaporation and Crystallization

1 Evaporation involves concentrating:
    A __A liquid in a solid
    B **X** A solid in a liquid
    C __A gas in a liquid
    D __A liquid in a gas
    Answer: (b) Evaporation is the removal, either by heat input or vacuum, of a solvent in which a solid is dissolved in. This process increases the concentration of the non-volatile material.

2 The primary design equation for an evaporator considers:
    A __Boiling point of the liquid
    B __Pressure in the evaporator
    C __Temperature difference between the heat source and the boiling point of the solution
    D **X** All of the above
    Answer: (d) The design equation is basically Q (energy required) = U (heat transfer coefficient) $\times A$ (area available) $\times \Delta T$ (temperature difference). The boiling point of the solution is part of the ΔT calculation; the pressure affects the boiling point; and the temperature difference is the ΔT.

3 The boiling point of a solution to be evaporated with steam is affected by all but:
    A __Pressure
    B **X** Steam price
    C __Concentration of dissolved salts
    D __Temperature differential between heat source and boiling point of the solution
    Answer: (b) The steam price may affect the cost of the evaporation step, but not technical aspects as mentioned in the other three choices.

4 The boiling point of a salt solution will ____ with increased concentration of the dissolved salt:

A __Decrease

B __Need more information to answer

C X Increase

D __Rise by the square root of concentration change

Answer: (c) Any increase in salt concentration will increase the boiling point. (d) is not correct as the relationship is not the same for all salts.

5 If the steam pressure feeding an evaporator slowly decreases with time, the salt concentration leaving the evaporator will ____over the same time period:

A __Increase

B X Decrease

C __Stay the same

D __Depends (on what?) ____

Answer: (b) If the steam pressure drops, its temperature will also drop, lowering the ΔT driving force.

6 Salt solution carry over into the vapor phase of an evaporator can be minimized through the use of:

A __Prayer

B __Filters

C __Cyclones

D X Demisters

Answer: (d) Demisters are usually used to accomplish this. A cyclone is the type of equipment more often used to remove dry particles from gas streams.

7 Multi-effect evaporators function by:

A X Using the vapor from one stage to vaporize another stage

B __Using the "super-effect" of steam

C __Condensing the first stage vapor and then boiling it a second time

D __Taking advantage of off-peak power prices

Answer: (a) Vapor driven off one stage of evaporation is used as the energy source to boil the solution in a second stage. This requires either pressure or vacuum as a driving force.

8 Film evaporators are used primarily for:

A X Temperature-sensitive and high viscosity materials

B __Emotionally sensitive materials

C __Hold your temper materials

D __All of the above

Answer: (a) Film evaporators (typically wiped continuously in some fashion) are used in these cases to minimize residence time in the evaporator.

9 The basic difference between evaporation and crystallization is that the solution is concentrated by:

A __Use of diamonds

B X Cooling

C __Any type of heat sources that is available except steam

D __Suction

Answer: (b) Though cooling can be used to evaporate under vacuum, crystallization most often involves cooling of a solution to precipitate out solids in solution to recover them for further treatment.

10 The types of crystals produced in a crystallizer are affected by:

A __Phase diagram

B __Rate of cooling

C __Amount of agitation

D X All of the above

Answer: (d) Any of these parameters will affect the types of crystals produced. The phase diagram (if it exists) will define and put limits on the operating parameters that can be used.

11 A phase diagram for a salt and solvent will determine all but:

A __The types of crystals that will be formed

B __Where various hydrates will form as a function of temperature and concentration

C X Cost incurred to operate at a particular point within the phase diagram

D __How to produce certain types of salt hydrates

Answer: (c) The phase diagram, if it exists, will define where it is possible to operate, but says nothing directly about the cost of operating within it at any point.

Chapter 13 Review: Liquid–Solids Separation

1 The driving force for filtration is:

A X Pressure differential

B __Concentration differential

C __Temperature differential

D __Temperament differential

Answer: (a) Pressure differential is what forces the liquid through a medium, leaving solids behind.

2 A pre-coat on a filter medium may be required if:
  A __It is cold in the filter operations room
  B __The operating instructions say so
  C __The particle size of the solids is greater than the hole size in the filter medium
  D X_The particle size of the solids is smaller than the hole size in the filter medium
  Answer: (d) A particle size smaller than the pore size of filter medium can quickly plug the filter. A pre-coat of a large particle, inert material is typically used to provide a physical barrier to this happening.

3 If the filtration is run under constant pressure, the flow rate will ____ with time:
  A X_Drop
  B __Stay the same
  C __Increase
  D __Need more information to know
  Answer: (a) At constant pressure, the bed of filtered solids will increase, providing additional pressure drop. If the feed pressure is not increased, the flow of liquid exiting the filter will decrease.

4 If the filtration is run to produce constant volume output, the pressure will ____with time:
  A __Drop
  B __Stay the same
  C X_Increase
  D __Rise by the cube of the change in flow
  Answer: (c) This is the opposite situation compared to question 3. In order to maintain a constant flow of filtrate, the pressure (or differential pressure) will need to be increased.

5 Raising the solids concentration in a filter feed, with other variables unchanged, will ____the filtration rate:
  A __Increase
  B X_Decrease
  C __Not affect
  D __Drop by the square root of the solids concentration change
  Answer: (b) Increased solids will increase filter cake volume, thus increasing pressure drop and decreasing filtrate rate.

6 Increased compressibility of a filtration cake will _____the filtration rate over time:

A __Increase

B X Decrease

C __Stay the same

D __Increase by the square of the compressibility

Answer: (b) A higher compressibility will decrease the void volume between solids, reducing the filtrate rate over time.

7 A centrifuge adds what force to enhance filtration rate:

A __Gravity

B __Pressure

C X Centrifugal/centripetal

D __Desire for a faster rate

Answer: (c) Rotational force is added to the existing pressure differential.

8 The rate of filtration in a centrifuge is proportional to the _____of the rotational speed:

A __Linearity

B __Square root

C __Cube

D X Square

Answer: (d) This is similar to any other rotational device's response to rotational speed.

Chapter 14 Review: Drying

1 Drying is defined as the removal of ___from a solid material:

A X Solvent

B __Coolant

C __Water

D __Spirits

Answer: (a) The solvent might be water, but not always!

2 Drying rate is affected by all but:

A __Solvent concentration at any point in time

B X Cost of vacuum or steam

C __Agitation within the dryer

D __Temperature difference between solid and heating medium

Answer: (b) The other three variables can all affect the rate of drying, but the cost of the drying energy source only affects the cost, not the rate.

3 Key variables in the design and operation of a spray dryer include:

A __Liquid or slurry to gas ratio

B __Viscosity of fluid and pressure drop across the spray nozzle

C __Temperature difference between hot drying gas and liquid

D X All of the above

Answer: (d) All of these process and physical property values affect the design and operation of a spray dryer.

4 Design issues with rotary dryers include:

A __Possible need for dust recovery

B __Particle size degradation

C __Dust fires and explosions

D X All of the above

Answer: (d) All of these are important, especially concerns about dust fires and explosions. Since this type of dryer can reduce particle size through attrition, it can enhance the probability of dust explosions.

5 Freeze drying is a potential practical drying process if:

A __A freezer is available

B X The S–L–V phase diagram allows direct sublimation at a reasonable vacuum

C __It is desired to have a cold product

D __The plant manager owns stock in a freeze dryer manufacturing company

Answer: (b) We must know the basic solid–liquid–vapor phase diagram to be able to judge the feasibility and cost of a freeze drying process.

6 A drying rate curve tells us:

A __How fast a solid will dry

B __How much it will cost to dry a solid to a particular residual water or solvent level

C X The drying rate curve as a function of residual water or solvent

D __How the cost of drying is affected by the rate of inflation

Answer: (c) A drying rate curve shows the residual moisture or solvent level as a function of time under any given process condition. This information, along with the cost of any drying process or equipment, is used to determine options on a drying process.

7 Auxiliary equipment frequently needed for a drying process include:

A X Cyclones and scrubbers

B __Backup feed supply

C __Customer to purchase product

D __Method of measuring the supply chain

Answer: (a) Dryers frequently generate dust as particle size is reduced through attrition. This often requires the use of cyclones and/or scrubbers to minimize the loss and discharge of solid particles into the atmosphere. There may also be environmental regulations requiring such dust control measures.

Chapter 15 Review: Solids Handling

1 The energy used in particle size reduction is primarily a function of:
A __Price of energy
B X Ratio of incoming particle size to exiting particle size
C __Size of the hammers or pulverizers
D __Strength of the operator running the equipment
Answer: (b) The ratio of size reduction is the key variable. The greater this ratio is, the more energy is required.

2 Cyclones have primarily one very positive design feature and one very negative design feature:
A X No moving parts and sharp cutoff in particle separation
B __No motor and low particle collection
C __Can be made in the farm belt but cannot collect large size corn cobs
D __Are small and can make a lot of noise
Answer: (a): They have no moving parts (they use impact and geometry), but their capabilities are limited in terms of sharpness of separation. It is critical to match performance with expectations through testing.

3 A key solids characteristic in assessing solids cohesion is:
A __Particle size
B __Height of solids pile
C __Slope of laziness
D X Angle of repose
Answer: (d) The angle of repose is an indirect measurement of particle cohesive strength. The greater is it, the more difficult it is for particles to flow in a storage situation.

4 A poorly designed hopper can cause:
A __No flow when the bottom valve is opened
B __Segregated particle size flow
C __Surges in flow behavior
D X All of the above
Answer: (d) All of these things are possible results from a poorly designed hopper.

Chapter 16 Review: Tanks, Vessels, and Special Reaction Systems

1 Simple storage tanks can be hazardous due to:
A __Leakage
B __Overfilling and under filling
C __Contamination
D X All of the above
Answer: (d) Any of these conditions are cause for concern. Leakage and contamination may be longer term or quality issues, but under filling and over filling can be catastrophic in terms of consequences.

2 A key design feature in an agitated vessel (its Z/T ratio) is:
A X Its height-to-diameter ratio
B __Which company manufactured it
C __The engineer who designed it
D __When it was placed into service
Answer: (a) The height to diameter ratio is critical in terms of how effective the mixing and agitation is both top to bottom and side to side. The physical properties of the material are also critical in terms of the difficulty in achieving uniformity.

3 Choosing an agitator for a tank will be affected by:
A __Densities of liquids and solids
B __Viscosity of liquids
C __Ratios of gases/liquids/solids
D X All of the above
Answer: (d) Every one of these variables will have a significant effect on agitation efficiency and energy use. It is important to measure these variables at the temperature and pressure conditions that will actually exist.

4 The significance of top-to-bottom agitation will be affected by:
A __Gas and liquid densities and their changes over time
B __Formation of solids as a reaction proceeds
C __Necessity for liquid/solid/gas mixing
D X Any of the above
Answer: (d) All of these are important variables and the relative importance of each will determine optimum tank geometry. It is useful to envision extremes such as a gas with a slow reaction in a viscous solvent reacting with a solid slurry.

5   Shaft horsepower requirements for an agitated vessel will be MOST affected by:

A ___Air flow available and ability to clean an exiting air stream

B ___Physical properties of materials being mixed and agitated

C ___Particle size degradation

D X All of the above

Answer: (d): As with question #5, all of these can be important and the degree to which each is important will be a function of the vessel's function and the physical property differences.

Chapter 17 Review: Chemical Engineering in Polymer Manufacture and Processing

1   Polymers are long chains of:

A ___Polys

B ___Mers

C ___Various mixtures of polys and mers

D X Monomers

Answer: (d) Polymers are long chains of monomers (ethylene, styrene, etc.) whose double bonds have been activated is such a manner that they can react with each other to produce long chains.

2   Latex polymers are unique in that they:

A ___Are used in tennis shoes

B ___Are stretchable

C X Are suspensions of polymers in solution

D ___Are combinations of thermoplastic and thermoset polymers

Answer: (c) Latexes are not actually polymers themselves, but suspensions of polymers (via the use of surface chemistry on the polymer particles). When applied, for example as paints, the water/solvent evaporates, leaving a polymer film behind.

3   The uniqueness of elastomers is that they_____

A X Have a T_g below room temperature

B ___Have a T_g at room temperature

C ___Have a T_g above room temperature

D ___Have a T_g that is controllable

Answer: (a) Elastomers have a glass transition temperature (where the polymer is not rigid) below room temperature. A conventional rubber band is an example of such a polymer.

4 *cis-* and *trans-*isomers differ in:
  A X Where functional groups are positioned on the monomer backbone
  B __Their preference for being *cis* or *trans*
  C __Their ability to change positions
  D __Cost
  Answer: (a) *Cis* and *trans* refer to whether a functional group is positioned on only one side or alternative sides of a polymeric double bond. This positioning can greatly affect physical properties.

5 "Condensation" polymerizations are unique in that they:
  A __Produce polymers that condense when it is cold
  B X Split out a molecule during the polymerization process
  C __Prevent another monomer from entering the process
  D __Provide a barrier to another polymerization process occurring
  Answer: (b) This type of polymerization involves chemical reactions as well as the joining of monomers. Typically a molecule such as water or hydrochloric acid (HCl) is produced when two different monomers are polymerized. This allows new composition polymers to be produced, but it also adds the complication of removing the "split out" molecule from the desired product and the process.

6 Polymer additives can be used to affect or change:
  A __Color
  B __Flowability
  C __Ability to foam
  D X All of the above
  Answer: (d) Polymer capability and appearance can be enhanced or modified in any of these areas through the addition of additives or latent properties (i.e., foaming) later activated by the user.

7 A differential scanning calorimetry (DSC) tells us:
  A __Softening temperature range of a polymer
  B __Softening temperature ranges within a polymer
  C __Degree of crystallinity within a polymer
  D X All of the above
  Answer: (d) A well run DSC will tell us all of the above properties, most of which are related to the temperature of use of the polymer.

8 Chemical engineering challenges presented in processing polymers include all but:
  A __High viscosity
  B __Slow heat transfer rates
  C X Knowledge of the polymers being processed
  D __Difficulty in mixing
  Answer: (c) Knowledge of the polymer name or type is not a chemical engineering challenge; it is a basic piece of knowledge. The chemical engineering challenges are the other three process and product variables.

9   The technical challenges in recycling of plastics and polymers can be affected by all but:
  A __Purity of plastics and ability to separate into various types
  B __Energy value of plastics
  C X Legislation
  D __Landfill availability
  Answer: (a) Unfortunately, legislation (especially that based on politics and not technical information) does not change the nature of the technical challenges represented in the other three answers.

Chapter 18 Review: Process Control

1   Proper process control is necessary because:
  A __Specifications must be met on products produced
  B __Environmental emissions must be within permitted limits
  C __Safety and reactive chemicals issues must be controlled
  D X All of the above
  Answer: (d) All of these aspects are important. The ranking of them may vary from process to process.

2   The elements of a process control loop include all but:
  A __Method to measure
  B X Manager's approval
  C __Way to evaluate the measurement versus what is desired
  D __Corrective action
  Answer: (b) The manager's approval has nothing to do with what is required! Hopefully the manager understands the decisions on control loop architecture and why they are the way they are.

3   Characteristics of a process which affect how it should be controlled include all but:
  A __Customer quality requirements
  B __Response time of measurements
  C __Degree of deviation permitted around the set point
  D X The mood of the process operator that day
  Answer: (d) The process must be controlled according to these specifications no matter the mood of anyone in the control room!

4   The least sophisticated process control strategy is:
  A X On–off
  B __Off–on
  C __Off sometimes, on other times
  D __On–on, off–off
  Answer: (a) The control system turns on the "final control element" when the measurement deviates, then turns off when it again reaches the set point desired.

5 Integral control has a key advantage in that it:
   A __Has the capability to integrate
   B X Will eventually result in the process reaching its desired set point
   C __Oscillates around the desired set point
   D __Oscillates in a controlled manner around the desired set point
   Answer: (b) The controlled value may oscillate in a slowly degrading manner for a while, but the integrating action will eventually bring the controlled variable to its desired set point. Settings on the integral loop can be used to control the degree of oscillation and the time to reach the desired value.

6 Derivative control allows a control system to:
   A X Anticipate a change in process based on input changes
   B __Preplan batch operations
   C __Provide a more uniform break structure for process operators
   D __React sooner to post process changes
   Answer: (a) If it is known how a process will respond to an input change, then derivative control can be used to anticipate the change by making control element settings prior to the downstream change being measured.

7 If a process has a slow reaction rate process feeding into a very fast final reaction process, the type of process control likely to be used is:
   A __Proportional
   B __Follow the leader
   C __Wait 'til it tells me
   D X Cascade
   Answer: (d) One control loop design will be incorporated within another to be appropriate to the particular process.

8 Control valves can be characterized, in terms of their process response, by all but:
   A __Capacity
   B __Speed of response
   C __Type of response
   D X Materials of construction
   Answer: (d) Materials of construction will have little or no effect on how the valve responds. They will certainly affect costs, weight, etc.

9 A control valve curve plots:
   A __% Open versus % closed
   B __% Closed versus flow rate
   C X % Flow versus % open
   D __Flow rate versus air pressure supplied
   Answer: (c) The other plots may be useful to know, but they are not the standard method of plotting control valve response.

10 A water cooling control valve, with loss of utilities, should fail to open if:

    A __An endothermic reaction is being run

    B <u>X</u> An exothermic reaction is being run

    C __The utility water rates are temporarily dropping

    D __There is no mechanic available to close

Answer: (b) Since an exothermic reaction is releasing heat, it is critical that any cooling system fail in a way to maximize cooling rate.

11 A control valve, with loss of utilities, should fail to close if:

    A __An endothermic reaction is being run

    B <u>X</u> It controls feed to an exothermic reaction

    C __There is one person strong enough to close it

    D __There is no other option

Answer: (b) A feed to an exothermic (heat producing) reaction must be closed with loss of utilities to prevent further heat generation.

12 A control room variable may not indicate the actual process conditions if:

    A __The sensor has failed or been disconnected

    B __A physical property effect has not been taken into account

    C __The operator is not looking at the correct screen

    D <u>X</u> Any of the above

Answer: (d) Any of these can cause a control room screen or data display to be incorrect. It is imperative that a combination of operating discipline, layers of protection, and HAZOP reviews ensure that this possibility is minimized.

Index

Chemical Engineering for Non-Chemical Engineers, First Edition. Jack Hipple.
© 2017 American Institute of Chemical Engineers, Inc. Published 2017 by John Wiley & Sons, Inc.

Printed and bound by CPI Group (UK) Ltd, Croydon, CR0 4YY

08/03/2023

03200083-0001